Marine Protected Areas

Marine Protected Areas

Principles and techniques for management

Edited by Susan Gubbay
Environmental Consultant
Ross-on-Wye
UK

CHAPMAN & HALL
London · Glasgow · Weinheim · New York · Tokyo · Melbourne · Madras

Published by Chapman & Hall, 2–6 Boundary Row, London SE1 8HN

Chapman & Hall, 2–6 Boundary Row, London SE1 8HN, UK

Blackie Academic & Professional, Wester Cleddens Road, Bishopbriggs, Glasgow G64 2NZ, UK

Chapman & Hall GmbH, Pappelallee 3, 69469 Weinheim, Germany

Chapman & Hall USA, 115 5th Avenue, New York NY 10003, USA

Chapman & Hall Japan, ITP-Japan, Kyowa Building, 3F, 2-2-1 Hirakawacho, Chiyoda-Ku, Tokyo 102, Japan

Chapman & Hall Australia, Thomas Nelson Australia, 102 Dodds Street, South Melbourne, Victoria 3205, Australia

Chapman & Hall India, R. Seshadri, 32 Second Main Road, CIT East, Madras 600 035, India

First edition 1995

Typeset in 10/12pt Sabon by Acorn Bookwork, Salisbury, Wiltshire
Printed in Great Britain by St Edmundsbury Press Ltd, Bury St Edmunds, Suffolk

ISBN 0 412 59450 1

A catalogue record for this book is available from the British Library

∞ Printed on acid-free text paper, manufactured in accordance with ANSI/NISO Z39.48-1992 and ANSI/NISO Z39.48-1984

Contents

Contributors

Mr Billy Causey
Sanctuary Superintendent
Florida Keys National Marine Sanctuary
PO Box 500368
Marathon, FL 33050
USA

Dr John Gibson
Centre for Marine Law and Policy
Cardiff Law School
University of Walcs
PO Box 427
Cardiff CF1 1XD
UK

Dr Susan Gubbay
Environmental Consultant
55 Verschoyle Gardens
Ross-on-Wye HR9 7HH
UK

Dr Stephanie Kaza
UVM Environmental Program
University of Vermont
153 South Prospect Street
Burlington
Vermont 05401
USA

Mr Graeme Kelleher
Great Barrier Reef Marine Park Authority
PO Box 791
Canberra ACT 2601
Australia

Mr Richard Kenchington
Great Barrier Reef Marine Park Authority
PO Box 791
Canberra ACT 2601
Australia

Dr Dan Laffoley
Marine Task Force
English Nature
Northminster House
Peterborough PE1 1UA
UK

Dr Andrew Price
Ecosystems Analysis and Management Group
Department of Biological Sciences
University of Warwick
Coventry CV4 7AL
UK

Dr Rod Salm
IUCN
East Africa Regional Office
PO Box 68200
Nairobi
Kenya

Dr Dermot Smyth
Consultant in Cultural Ecology
PO Box 1202
Atherton Qld 4883
Australia

Dr Lynda Warren
Centre for Marine Law and Policy
Cardiff Law School
University of Wales
PO Box 427
Cardiff CF1 1XD
UK

Ms Susan Wells
University College of Belize
PO Box 990
Belize City
Belize

Ms Sarah Welton
Marine Conservation Society
38 West Street
Bere Regis
Wareham, Dorset BH20 7HW
UK

Dr Alan White
Coastal Resource Center
University of Rhode Island
Narragansett Bay Campus
Narragansett, RI 02882
USA

Preface

Marine protected areas (MPAs) have an important role in marine conservation programmes around the world. Although most have been established relatively recently when compared with protected areas on land, there is considerable expertise on their identification, setting up and management. Some techniques have been adapted from those used on land. Others are novel, and unique to marine conservation. The chapters in this book give an insight into this fast developing field where experiment and innovation work alongside techniques which have been tried and tested. The guiding principles behind key stages in the setting up and management of MPAs are described, and case studies illustrate how they have worked. While it is most encouraging to read about the successes, the case studies also point to difficulties which have been encountered. Not all of the examples are new or recent but, together, they illustrate what is happening in this field.

The guide for MPA planners and managers, edited by Rodney Salm and John Clark, based on workshops at the World Congress on National Parks in 1982, was an important landmark in this field when it was published in 1984. During the 1982 Congress, only a small group attended the sessions on MPAs. In 1992, several hundred people took part in the discussions on the subject, demonstrating the tremendous growth in effort and increase in interest in MPAs. I hope this publication will help that effort by encouraging, supporting and informing those involved with marine protected areas in particular and marine conservation in general.

Acknowledgements

My thanks are given to the authors for the time and effort they put into preparing chapters for this book. Their practical experience and insight into various aspects of marine protected areas have been invaluable and are what made this book possible. I am also extremely grateful to them for their patience with a project that took longer than expected to reach its conclusion.

Many thanks also to my husband, Alan Davis, for his encouragement throughout the project and assistance with the final manuscript. His support and patience were much appreciated.

1

Marine protected areas – past, present and future

Susan Gubbay

Marine protected areas have become the flagships of marine conservation programmes in many parts of the world. For the public and politicians, they are often the most tangible part of a marine conservation programme – it is possible to visit a marine protected area, to see what is being done to promote conservation at the site, and to be a part of its success or failure through individual actions. For the conservation manager, they are an opportunity to concentrate effort and resources on protecting marine wildlife and habitats. Action taken at these sites also makes a contribution to wider conservation efforts. Two very different examples of this are their potential as a reservoir for species which can seed other areas, and as a good base for marine education programmes. Marine protected areas often get a lot of public attention and they are at the leading edge of the marine conservation programmes in many countries.

1.1 A HISTORICAL PERSPECTIVE

The world's first marine protected area (MPA) was probably the Fort Jefferson National Monument in Florida, which covered 18 850 ha of sea and 35 ha of coastal land. Although this site was designated in 1935, the main impetus for MPAs came much later. The World Congress on National Parks in 1962 was one of the first international conservation meetings to give the subject special attention; a follow-up meeting, in 1982, called for the incorporation of marine, coastal and freshwater sites into the worldwide network of protected areas (IUCN, 1987). More recently, the Caracas Action Plan (IUCN, 1994a), from the IV World Congress on National Parks and Protected Areas in 1992, promoted four objectives:

1. The integration of protected areas into larger planning frameworks.
2. Expanding support for protected areas.
3. Strengthening the capacity to manage protected areas.

Marine Protected Areas: principles and techniques for management. Edited by Susan Gubbay.
Published in 1995 by Chapman & Hall, London. ISBN 0 412 59450 1.

4. Expanding international cooperation in the financing, development and management of protected areas.

All of these are relevant to MPAs but there were several actions specific to MPAs. Action 3.5 of the Caracas Action Plan ('Give attention to the special requirements for managing marine protected areas') lists the following:

- Contribute to a global system for categorizing coastal marine regions as the basis for assessing the adequacy of protected areas in these regions.
- Participate actively in coastal zone management programmes and ensure that both marine and terrestrial protected areas are used as key management tools in such programmes.
- Develop and implement integrated management programmes for marine protected areas.

Growing support for MPAs has come from other international initiatives as well. The Convention on Wetlands of International Importance Especially as Waterfowl Habitats, also known as the Ramsar Convention, requires parties to designate at least one wetland in their territories for the Ramsar List of Wetlands of International Importance. The definition of a wetland used by the Convention includes 'areas of marine water, the depth of which at low tides does not exceed six metres' (Article 1.1). The site selection criteria include guidance on assessing sites for their value as representative or unique wetlands, their importance to plants and animals and, more specifically, their importance to waterfowl. This has enabled the designation of MPAs under the Ramsar Convention.

Wider ranging conservation agendas developed by the international community have also been promoting MPAs as a critical part of marine conservation programmes. This can be seen in the report of the Bruntland Commission in 1987 (World Commission on Environment & Development, 1987) through to 'Agenda 21', of the United Nations Conference on Environment and Development in 1992.

The higher policy profile for MPAs is being translated into action through protected area programmes around the world. Initially, MPAs were often small extensions of terrestrial protected areas and were only included for ease of drawing boundaries. A bay, for example, would be included if the boundary was a direct line from one headland to the next. In these situations, little attention was given to the marine section of the protected area. As support for MPAs has grown, marine sites have become established in their own right. The majority are adjacent to the coast or close inshore where there is greatest pressure on marine systems from human activity, but MPAs can cover whole seas. The Indian Ocean whale sanctuary, where controls apply to a single group of animals, is one example.

1.2 TYPES OF MARINE PROTECTED AREAS

The following definition of marine protected areas was developed at the 4th World Wilderness Congress and adopted by IUCN at its 17th General Assembly in 1988:

> Any area of intertidal or subtidal terrain, together with its overlying water and associated flora, fauna, historical and cultural features, which has been reserved by law or other effective means to protect part or all of the enclosed environment.

Sites which fit this definition have been given a variety of names. In common with protected areas on land they may be called 'reserves', 'sanctuaries', 'parks', or have some other title. To clarify the situation and guide protected area managers, the IUCN, through its Commission on National Parks and Protected Areas, has developed and refined a list of the categories of protected area. Ten categories were identified in 1978 and, following the most recent review, the list has been reduced to six. These are areas managed mainly for:

- I Strict protection (i.e. Strict Nature Reserve/Wilderness Area)
- II Ecosystem conservation and recreation (i.e. National Park)
- III Conservation of natural features (i.e. Natural Monument)
- IV Conservation through active management (i.e. Habitat/Species Management Area)
- V Landscape/seascape conservation and recreation (i.e. Protected Landscape/Seascape)
- VI Sustainable use of natural ecosystems (i.e. Managed Resource Protected Area).

They can be distinguished on the basis of the objectives of management (IUCN, 1994b) (Table 1.1) and there are marine examples of each type (Table 1.2). Some specific objectives of a global representative system of MPAs were also adopted by the IUCN in a resolution of their 17th General Assembly in 1988. These were:

- to protect and manage substantial examples of marine and estuarine systems to ensure their long-term viability and to maintain genetic diversity;
- to protect depleted, threatened, rare or endangered species and populations and, in particular, to preserve habitats considered critical for the survival of such species;
- to protect and manage areas of significance to the life cycles of economically important species;
- to prevent outside activities from detrimentally affecting the marine protected areas;

Table 1.1 Matrix of management objectives and IUCN protected area management categories

Management objective	*Ia*	*Ib*	*II*	*III*	*IV*	*V*	*VI*
Scientific research	1	3	2	2	2	2	3
Wilderness protection	2	1	2	3	3	–	2
Preservation of species and genetic diversity	1	2	1	1	1	2	1
Maintenance of environmental services	2	1	1	–	1	2	1
Protection of specific natural/cultural features	–	–	2	1	3	1	3
Tourism and recreation	–	2	1	1	3	1	3
Education	–	–	2	2	2	2	3
Sustainable use of resources from natural ecosystems	–	3	3	–	2	2	1
Maintenance of cultural/traditional attributes	–	–	–	–	–	1	2

Key 1 Primary objective
2 Secondary objective
3 Potentially applicable objective
– Not applicable

Table 1.2 Examples of MPAs in each of the IUCN protected area categories (information from IUCN, 1994c)

Category	*Type*	*Marine examples*
I	Strict protection	Laut Banda, Indonesia Leigh Marine Reserve, New Zealand
II	Ecosystem conservation and recreation	Ras Mohammed National Park, Egypt Hol Chan Marine Nature Reserve, Belize Sandy Bay, Honduras Watamu, Kenya Sanganeb Atoll, Sudan
III	Conservation of natural features	USS *Monitor* NMS, USA
IV	Conservation through active management	Galapagos Marine Reserve, Ecuador Sangalaki, Indonesia Isole Tremiti, Italy Skomer, UK
V	Landscape/seascape conservation and recreation	Côte Bleue, France Northern Sporades, Greece Taklong Island, Philippines Cordell Bank, USA
VI	Sustainable use of natural ecosystems	Kiunga Marine National Reserve, Kenya

- to provide for the continued welfare of people affected by the creation of marine protected areas; to preserve, protect, and manage historical and cultural sites and natural aesthetic values of marine and estuarine areas, for present and future generations;
- to facilitate the interpretation of marine and estuarine systems for the purposes of conservation, education and tourism;
- to accommodate with appropriate management regimes a broad spectrum of human activities compatible with the primary goal in marine and estuarine settings;
- to provide for research and training, and for monitoring the environmental effects of human activities, including the direct and indirect effects of development and adjacent land-use practices.

The variety of objectives and types of MPAs makes comparisons difficult. Some are designed as strictly protected areas while others are concerned with multiple-use management. The total number of MPAs is also difficult to establish because there are many ways in which the count can be made. Sites with a marine component, however small, could be included or, alternatively, only protected areas which are predominantly marine. Fisheries management areas or general marine management programmes applied to whole seas may be included in some cases and not in others. DeSilva *et al.* (1986) identified 430 MPAs and the most recent estimate (1994) is 1111 MPAs (Table 1.3) (Kelleher *et al.*, in press). They range in size from the Great Barrier Reef Marine Park, which covers around 344 000 km^2, to MPAs of a few square kilometres.

1.3 THE MANAGEMENT OF MPAS

The approach to management of MPAs varies depending on the type of site, the aims and the circumstances under which protection is being achieved. The lead organization may be a local community group, an educational establishment, a conservation organization or a government agency; the controls may be voluntary or backed through statutory powers; and the protected habitats, communities and species range from polar to tropical.

Despite this variety, some consistent approaches emerge. The chapters in this book give an insight into this fast developing field where experiment and innovation work alongside techniques which have been tried and tested. The guiding principles behind key stages in the setting up and management of MPAs are described, and case studies illustrate how they work. While it is most encouraging to read about the successes, the case studies also point to difficulties which have been encountered. Not all of the examples are new or recent but, together, they illustrate what is happening in this field. The

Table 1.3 Marine protected areas (from Kelleher *et al.*, in press)

Region	*Number of MPAs*
Antarctic	24
Arctic	9
Mediterranean	46
North-West Atlantic	45
North-East Atlantic	34
North Atlantic – Baltic	24
Wider Caribbean	76
West Africa	25
South Atlantic	25
Indian Ocean	25
North-West Indian Ocean	24
South-East Africa	56
South-East Asia	65
Central and South Pacific	49
North-East Pacific	54
North-West Pacific	201
South-West Pacific	25
Australia	304
Total	1111

strongest theme is the importance of ensuring that people understand and are involved in the establishment and management of MPAs. This is fundamental to the successful implementation of management plans, zoning arrangements, enforcement and other activities relating to MPAs. Education and interpretation programmes help to achieve this understanding and are seen as a vital part of most MPA programmes. Experience to date has been usefully collated into guides for planners and managers (e.g. Ray, 1976; Salm and Clark, 1984; Kelleher and Kenchington, 1992) (see also Chapters 2 and 5) and a global overview of the sufficiency of the existing network of MPAs is currently under examination (Kelleher *et al.*, in press).

MPAs have succeded in a variety of ways. They have prevented further degradation of marine habitats and communities while enhancing resources both inside and outside their boundaries; they have involved the public in marine conservation and raised public awareness of marine issues of concern; they have been of economic benefit to the locality; and they have provided useful conditions for marine research. The transfer of knowledge about these benefits, and how they have been achieved, is important if MPA programmes are to improve. To help this process each of the following chapters includes case studies describing the actions taken at particular MPAs and how effective they have been.

1.4 FUTURE PROSPECTS FOR THE DEVELOPMENT OF MPAS

Marine protected areas are a well established part of most marine conservation programmes and are likely to remain so. The International Convention on the Conservation of Biodiversity 1992 has given added impetus to this approach by advocating the use of protected areas (Article 9). The World Bank, through the Global Environment Facility, is funding a study on the role of MPAs in the conservation of global marine biodiversity (Kelleher *et al.*, in press). In the near future the emphasis is likely to be on the establishment of more MPAs, as marine conservation catches up with conservation efforts on land. However, there is also recognition of the need to improve the effectiveness of existing MPAs. Views on the role of MPAs in conservation programmes are also developing and this will influence their future.

The following sections highlight issues which conservation managers are likely to consider as they identify the role of MPAs in the twenty-first century.

1.4.1 Site selection

The current focus on biodiversity conservation is an important influence on MPA programmes. This has implications for site selection but it remains unclear whether much will change. The types of area listed by Norse (1993) as especially important for marine biodiversity, for example, are the sorts of site which have been selected traditionally as MPAs:

- Areas of high diversity
- Areas of high endemism
- Areas of high productivity
- Spawning areas that serve as a source of recruits
- Nursery grounds
- Migration stopover points and bottlenecks.

In addition to having MPAs for 'special' places most MPA programmes include representative sites and these are likely to get greater attention in programmes directed at the conservation of marine biodiversity. The biodiversity debate should also lead to greater consideration of the processes which keep these systems functioning, and whether MPAs can make a contribution on this wider issue. One example is the work of Ray and McCormick-Ray (1992) who have listed a set of guidelines for establishing a representative system of marine and estuarine protected areas based on a systems approach, as part of the current debate on approaches to the conservation of biodiversity. These are:

Guideline 1 – Define operational objectives
Guideline 2 – Gain public support
Guideline 3 – Apply scientific procedures
Guideline 4 – Develop a central information system
Guideline 5 – Represent diversity of the land/seascape
Guideline 6 – Emphasize land–sea interactions
Guideline 7 – Relate objectives to appropriate scale
Guideline 8 – Focus on ecosystem processes
Guideline 9 – Set protected areas in a regional context
Guideline 10 – Attain a national network.

The establishment of MPAs to improve degraded environments has not been so well explored. In most cases managers try to select sites which are as 'pristine' as possible and try to safeguard them for the future. A good example of this is the Poor Knights Marine Reserve in New Zealand. On the other hand the Leigh Marine Reserve, also in New Zealand, was selected as a 'representative site' with little apparent damage before a no-take reserve was established. Despite this, significant changes were observed afterwards (Ballentine, 1991), suggesting that strictly protected MPAs may already be acting as restoration areas. In the future, establishment of MPAs which are selected for their potential to be 'restored' could usefully complement the network of MPAs set up to prevent the degradation of sites.

The majority of MPAs lie inshore and are usually adjacent to coastline. This is where pressures on the marine environment from human activity are greatest and where there are many productive and diverse marine communities. The Stellwagen Bank Marine Sanctuary in the United States is one of relatively few totally submerged offshore MPAs and the value and role of MPAs in these sorts of locality need to be examined further, since marine conservation problems are no longer confined to the coastal zone. Some attention has already been given to the idea of having MPAs in the high seas. In 1991 the National Oceanic and Atmospheric Administration organized a workshop to discuss the concept of 'Wild Ocean Reserves' to protect valued resources in international waters. This sort of event may become more frequent, especially with the enforcement of the Law of the Sea Convention in November 1994 which gives added marine responsibilities to coastal nations. If this is considered to be a useful way forward, there will be difficult management issues to address, especially in relation to the enforcement of regulations in more remote localities.

1.4.2 Legislation

The legal frameworks used to designate and manage MPAs are described in Chapter 3. The debate about how much legal backing is needed, and about the best ways to achieve support for MPAs, will undoubtedly continue. A less well explored issue is the scope for legally protected MPAs beyond territorial waters. The declaration of Exclusive Economic Zones (EEZs) has pushed national jurisdictions further from the shore and the Law of the Sea Convention has resulted in the duties and responsibilities of nations extending over even larger areas. Traditional freedoms are already being contained to some extent through voluntary agreement, such as the International Maritime Organization's identification of areas to be avoided by shipping. Measures such as these can be used to back up protection of MPAs in EEZs and international waters.

Existing MPAs in international waters tend to be limited to particular activities or specific groups of animals. The International Convention for the Prevention of Pollution from Ships 1973 identified a number of 'Special Areas' where there are additional controls on shipping activity. The Southern Ocean whale sanctuary, which was agreed by the International Whaling Commission in 1994, is an area where all commercial whaling is banned south of the 55th parallel in the Indian Ocean, the 60th parallel off Chile and below the 40th parallel in the rest of the southern ocean. These MPAs illustrate another important trend, which is for nations to work together to make their marine conservation programmes more effective and to codify these with international agreements. The UNEP Regional Seas Programme, which focuses on semi-enclosed seas, illustrates this approach. The Kingston Protocol to the Cartagena Convention for the Protection and Development of the Marine Environment of the Wider Caribbean Region provides a legal framework for the Caribbean Environmental Programme (Freestone, 1990). The Convention on the Protection of the North-East Atlantic 1992 is another example. Similar agreements may well cover most of the world's oceans in the future and, together with the Law of the Sea Convention, could provide a legal framework in which MPAs might be instituted for international waters.

1.4.3 Involving the community

The need for community involvement in MPAs cannot be stressed strongly enough. Members of the community can be involved in roles as varied as identifying sites, putting forward proposals, being involved in consultations and being responsible for the management of MPAs. Chapters 4, 8 and 10 look at this subject in detail. The strong message is that the management of MPAs is dependent on community support and that one of the best ways to achieve this is through genuine opportunities for involvement. Community-run MPAs, such as the voluntary marine conservation areas in the UK, should become more widespread, but the issue of community involvement will become more complex for MPAs covering large areas and lying further offshore.

1.4.4 Management of MPAs

Management plans

Management plans are critical to MPAs. Without them, a great deal of effort may be put into the designation of a MPA, with limited commitment to the long-term management of the area. Lack of finance is usually a major factor preventing their preparation, as funding for consultation and the designation of sites is often simpler to obtain than for day-to-day management. The issue of whether users should pay for access to MPAs to help to offset these costs is likely to become increasingly pertinent. This has already been considered for the Great Barrier Reef Marine Park (Green and Padma, 1991) and both here and at other sites fees from permits have been retained by the park to assist with management. In the case of the Looe Key and Key Largo National Marine Sanctuaries in Florida, fines and monies recovered as damages can also be used for the management of the sanctuaries (Chapter 7). The pressure on MPAs from visitors is a related issue which is likely to become increasingly problematic. Visitors can bring much needed income to the locality but, as with protected areas on land, pressure of numbers and inappropriate behaviour causes damage. Controls on recreational use are likely to become a more familiar part of MPA management.

Another important issue is the emphasis of management plans for MPAs. At the moment, a great deal of effort has to be put into dealing with current problems, but it is to be hoped that management plans will become more proactive. While it is difficult to predict what new issues are likely to emerge, it is important that management arrangements provide a suitable framework in which to achieve the aims of the protected area. Wherever possible opportunities should be taken to introduce controls before issues get out of hand. Areas set up as 'no-take' reserves should not have such problems but it is an important consideration in multiple-use MPAs. The experience of a significant increase in gill-netting in the Skomer Marine Nature Reserve in the UK illustrates this point. At the time of designation there was no gill-netting, hence no controls were put in place. Gill-netting is now a significant activity but it is very difficult to introduce controls at this stage. The 25-year strategic plan for the Great Barrier Reef World Heritage Area is one example where objectives and strategies for the future have been identified to make the process more forward looking (GBRMPA, 1994).

Enforcement

The enforcement of regulations is necessary for MPAs to succeed, and a lack of support for this aspect can undermine their credibility. Future technological developments may assist with enforcement, by improving monitoring aids, but a positive approach to enforcement, by educating users, should be taken wherever possible. The importance of involving the community has already been mentioned and is discussed further in Chapters 4, 9 and 10.

Education and interpretation

Education is through the provision of information about the value, interest, concerns and ways of managing our activities in the marine environment. Hearing about the success of local initiatives can have far-reaching consequences. Support for the MPA programme in New Zealand grew from the success of the Leigh Marine Reserve, and support for coastal zone management for a whole county grew from the Purbeck Marine Wildlife Reserve in the UK. To complement the groundswell of support from local initiatives there is also a need for marine programmes to be promoted at national and international levels. National and international campaigns to explain the need for conservation management and the value of MPAs as part of that management are promoted by organizations like the World Wide Fund for Nature and IUCN (the World Conservation Union) but there is also a need for governments to promote such actions.

Holistic view

There are many aspects to the management of MPAs and so, not surprisingly, the emphasis at any one site depends on the aims and objectives of the MPA, and on issues specific to that location. As networks of MPAs become established it will become increasingly important to take a holistic look at the process of site management and how it contributes to the overall MPA programme. Monitoring, detailed evaluations and external audits are some of the issues which could usefully have a higher profile in the future.

1.4.5 The role of marine protected areas

Because there are many different types of MPA (section 1.2), it is inevitable that they have different roles in marine conservation programmes. This is likely to remain the case while there are MPAs with objectives ranging from 'no-take' to multiple use, but there are a number of areas where the role of MPAs could be examined in more detail to help to guide future programmes of site selection and management.

Economic benefit

Putting a financial value on the environment and carrying out cost/benefit analysis are controversial issues in environmental management but they are being considered as part of the discussions on the role of MPAs.

A cost/benefit analysis for the Virgin Islands National Park, for example, showed considerable profits to the local economy (Dixon and Sherman, 1990). In other places, such as the Tabaca Island Marine Park in Spain, there have been improvements in the fisheries of surrounding areas (Sanchez-Lizaso, in press). The US Man and Biosphere National Committee is undertaking a project to examine the ecological and socio-economic impacts of alternative access management strategies in MPAs. This aims to study changes in the resources of the reserve under various levels of restriction at four trial sites and to examine socio-economic effects on local income, recreational experience, and the value put on the area by the local population (MAB, 1994). There are situations where MPAs are promoted as benefiting the local economy and, as such, it is helpful to try to quantify what these benefits are and what sort of management brings these benefits.

'No-take' and multiple-use areas

Conservation managers see MPAs as one of a number of ways to advance marine conservation. As a consequence, there has always been discussion about the role of MPAs in wider conservation programmes because there are many different types of MPAs – from 'no-take' reserves and reserves set up for particular interest group such as fisheries, to 'multiple-use management areas'. A conscious decision to introduce integrated multiple use management for large areas, with various types of protection within them, would set a useful framework for MPAs and marine conservation programmes in general. Thus the approach taken for the Great Barrier Reef Marine Park could usefully become the norm for territorial waters and, within it, MPAs would be identified as no-take areas, fisheries management zones, sites of archaeological importance, etc. The biosphere reserve concept is another model which illustrates this approach, and there are now examples where it is being applied to areas of sea (Price and Humphrey, 1993). The move towards larger units of management is also shown by the consideration being given to management of large marine ecosystems (LMEs). Sherman *et al.* (1990) define these systems as large regions that have unique bathymetry, hydrography and productivity and within which biological populations have adapted reproductive, growth and feeding strategies. The focus for management in these cases is on functional ecological units which may be very large. Although marine management at this scale may seem a huge task, this need not be the case. Territorial waters are the obvious starting point and they are already multiple-use areas. What is usually lacking is **integrated** management of the activities in territorial waters and any strongly protected areas for nature conservation within them. With such a combined approach the role of MPAs can be appreciated more clearly as core conservation areas in a wider programme of marine management.

1.5 CONCLUSIONS

There is still the major task to be undertaken in convincing people of the need for marine conservation in general and for MPAs in particular, but attention is increasingly being focused on the role of MPAs in marine resource management. As coastal nations extend their jurisdictions and traditional freedoms become limited, there is greater responsibility for management of activities in the marine environment. Integrated multiple-use management arrangements may yet cover the entire area under a country's jurisdiction, and core regions of MPAs within these larger areas will provide greater protection for specific interests (such as nature conservation or fisheries) by excluding all or some activities. Integrated coastal and sea use management, and conservation programmes developed with an understanding of large marine systems in mind, are essential contexts for the MPAs of the future. As for the MPAs themselves, it is to be hoped that they will not only help to safeguard the most interesting, threatened and valued sites, but also protect the familiar and commonplace.

REFERENCES

Ballentine, W.J. (1991) *Marine Reserves for New Zealand.* University of Auckland, Leigh Marine Laboratory Bulletin No.25, Warkworth, New Zealand, 196 pp.

DeSilva, M.E., Gateley, E.M. and Desilvestre, I. (1986) *A bibliographic listing of coastal and marine protected areas: a global survey*, Woods Hole Oceanog. Inst. Tech. Rept. WHOI-86-11.

Dixon, J.A. and Sherman, P.B. (1990) *Economics of Protected Areas. A New Look at Benefits and Costs*, Earthscan Publications Ltd, London, 234 pp.

Freestone, D. (1990) Specially Protected Areas and Wildlife in the Caribbean – The 1990 Kingston Protocol to the Cartagena Convention. *Int. J. Est & Coastal Law*, 5(4), 362–382.

GBRMPA (1994) *Keeping It Great. A 25 Year Strategic Plan for the Great Barrier Reef World Heritage Area 1994–2019.* Great Barrier Reef Marine Park Authority, Townsville, 64 pp.

Green, G. and Padma, L. (1991) *Charging users of the Great Barrier Reef Marine Park. A report to the Great Barrier Reef Marine Park Authority*, Australian Bureau of Agriculture and Resource Economics, 80 pp.

IUCN (1987) *The Bali Action Plan.* Report during the World Parks Congress, Oct. 11–22, 1982. Commission of National Parks and Protected Areas, Ottawa, 29 pp.

IUCN (1994a) *Caracas Action Plan*, IUCN Programme on Protected Areas/CNPPA, Gland, Switzerland, 8 pp.

IUCN (1994b) *Guidelines for Protected Area Management Categories*, IUCN Commission on National Parks and Protected Areas with the assistance of the World Conservation Monitoring Centre. IUCN, Gland, Switzerland, 259 pp.

IUCN (1994c) *1993 United Nations List of National Parks and Protected Areas.* Prepared by the World Conservation Monitoring Centre and the IUCN Commission on National Parks and Protected Areas, with the support of the United Nations Environment Programme. IUCN, Gland, Switzerland, 313 pp.

Kelleher, G. and Kenchington, R. (1992) *Guidelines for Establishing Marine Protected Areas*. A Marine Conservation and Development Report. IUCN, Gland, Switzerland, vii + 79 pp.

Kelleher, G., Wells, S. and Bleakley, C. (in press). *Priority Areas for A Global Representative System of Marine Protected Areas*. Draft report to the World Bank Environment Department.

MAB (1994) Marine and Coastal Ecosystem Directorate Begins Core Project. *US MAB Bulletin*, **18**(2), p. 5.

Norse, E. (ed) (1993) *Global Marine Biodiversity. A Strategy for Building Conservation into Decision Making*. Center for Marine Conservation, Washington DC, 383 pp.

Price, A. and Humphrey, S. (eds) (1993) *Application of the Biosphere Reserve Concept to Coastal Marine Areas*. Papers presented at the UNESCO/IUCN San Francisco Workshop of 14–20 August, 1989. A Marine Conservation and Development Report. IUCN, Gland, Switzerland, viii + 114 pp.

Ray, G.C. (1976) Critical marine habitats, in *Proceedings of an International Conference on Marine Parks and Reserves, Tokyo, Japan, 12–14 May*, IUCN Publication (n.s.) No.37, Gland, Switzerland, pp 15–64.

Ray, G.C. and McCormick-Ray, M. (1992) *Critical Habitats and Representative Systems in Marine Environments: Concepts and Procedures*. Paper for Workshop II.11. The Science of Conservation in the Coastal Zone. IV World Congress on National Parks and Protected Areas, Caracas, Venezuela.

Salm, R.V. and Clark, J.R. (1984) *Marine and Coastal Protected Areas: A Guide for Planners and Managers*, IUCN, Gland, Switzerland, xiii + 302 pp.

Sanchez-Lizaso, J.L. (in press) *Marine Protected Areas in Coastal Zone Management*. Paper presented at the Coastal Zone Management Conference, 3–6 Sept., Dublin.

Sherman, K., Alexander, L.M. and Gold, B.D. (eds) (1990) *Large Marine Ecosystems: Patterns, Process and Yields*, American Association for the Advancement of Science, Washington DC.

2

Selection of marine protected areas

Rod Salm and Andrew Price

The selection of sites for conservation management is only one of many elements in the building of a network of marine protected areas (MPAs). It requires goal identification, site survey and data collection, data analysis, and data synthesis and plan formulation, all of which apply to site selection as well as all other steps in a marine protection area programme (Salm and Clark, 1984). While it is preferable to choose MPAs through a system planning exercise, in real life selection is often determined by crisis – some high level of threat to a site that is considered important for any reason.

In the first instance, guiding policies are usually defined to help fulfil the overall goal of the planning exercise. These policies should include lists of social, economic, ecological, regional and pragmatic criteria that are applied to evaluate candidate sites for selection. In the latter case, urgency imposed by threat prompts action that often ends in protection of the site.

Selection of sites according to a well-laid plan which includes an overall and clearly understood goal and focused, practical criteria is preferable to any *ad hoc* means, such as by crisis. Crises are likely to arise at intervals, but conservation managers need to get ahead of them if selection is to proceed along orderly and rational lines.

Initiatives for establishing protected areas can also come from sources such as local governments or inhabitants. Sites protected as a result of these initiatives have high local significance, but may contribute relatively little to national conservation objectives. In such cases, they have not been identified as areas of national importance because they have not been selected by the application of national level criteria. Nonetheless, because they enjoy strong local support and may require little manpower, time and money to manage, they should be viewed as useful contributions to conservation. However, they should not be considered in the national system of MPAs unless they truly are of national importance. The danger is that these areas will be considered as substitutes for larger, more valuable sites that really do contribute significantly to national objectives for conservation and development.

Selection criteria can be used to evaluate and select from among inherited,

Marine Protected Areas: principles and techniques for management. Edited by Susan Gubbay.
Published in 1995 by Chapman & Hall, London. ISBN 0 412 59450 1.

traditional and locally established protected areas those that contribute to national conservation objectives.

In this chapter we propose some basic principles and list criteria that guide selection of marine protected areas. Guidelines for establishing MPAs are also provided by Kelleher and Kenchington (1992).

2.1 GUIDING PRINCIPLES FOR THE SELECTION OF MARINE PROTECTED AREAS

First principle: the needs and abilities of the country define the goal and scope of the MPA programme.

It is easier to list principles and criteria from the theoretical perspective than it is to apply them in the real world. Common sense, and sensitivity to the true needs and abilities of the country, and its people, should be applied to the selection of sites.

Second principle: the goal of the MPA programme is the foundation of the selection process.

It is extremely important to have a clearly defined and focused goal for the system of MPAs. For example, a goal may be to establish a system of conservation areas for endangered species. In this case, the planning process would focus on identification of critical marine habitats (Ray, 1976) associated with the life functions (breeding, feeding, migration areas) of the target species. Criteria would emphasize ecological, regional and pragmatic parameters, and play down social and economic ones (section 2.3). If the goal were to protect critical habitats of fishery species, the selection criteria would include social and economic parameters and possibly place less importance on regional ones. An integrated goal that includes the needs of all sectors of society (e.g. fisheries, recreation, tourism) as well as those of the environment (wildlife, productive marine habitats, ecological process) is more ambitious, but may be easier to build into central development planning.

Third principle: the scope of the MPA programme (national, provincial, numbers and size of MPAs) defines the limits of the selection process and is an important determinant of the selection criteria.

It is increasingly difficult to select small areas in isolation from their surroundings, especially in the sea where winds, currents and species movements ensure a great deal of linkage between far distant areas. For this reason management of these small hot spots of activity needs to be built into that of the larger system of which they are a part (Salm and Dobbin, 1993). Establishment of MPAs in the broader context of coastal zone management planning has been recommended as an effective technique for management of upstream and other interactive processes and activities (Salm and Clark, 1984; IUCN/MEPA 1987a,b,c; Salm 1987). Coastal zone management plans can achieve a vast multiple-use reserve for the entire coastal zone of a nation. Within the coastal zone all valuable resource areas (critical marine habitats) can be managed to avoid damaging activities and pollution through development policies controlled by specific legislation or contained by zoning plans (Salm and Dobbin, 1987, 1989, 1993) (see also Chapters 3, 6 and 7).

Fourth principle: because each nation has different social, political, economic and environmental parameters, there can be no one definitive model for site selection.

Past experience has resulted in the evolution of a planning process with sets of criteria that can be adapted as appropriate to individual cases. The application of these criteria forms the main thrust of this chapter.

2.2 THE SITE SELECTION PROCESS

While it is difficult to isolate any components of the system planning process, four steps can be identified as essential to site selection. These are the collection, analysis, and synthesis of data leading to the identification of candidate sites, followed by the application of criteria to select specific sites for protection (Salm and Clark, 1984).

2.2.1 Data collection

Data collection from all available sources, including literature searches, interviews and field studies, is the first step in the identification and selection of MPAs. It is likely that existing data are incomplete. There may be detailed information for a small part of the country, and sparse or no data for the rest. A detailed plan for a series of field surveys will be required to provide essential social, biological and ecological information.

Data should be collected, evaluated for consistency and adequacy, and stored in easily retrievable form for analysis, such as on maps and in custom designed computerized databases that are linked by coordinates to the maps (Price, 1990).

2.2.2 Data analysis

Once data collection is complete, the information is analysed to show areas with concentrations of resources, human activities and threats to resources.

Other information, such as areas of conflicts among activities and other interests, or sites of specific interest (e.g. fish spawning grounds), can be extracted. Areas with concentrations of resources are all obvious candidates for protection. The specific objectives of the system planning process may, however, call for protection of specific sites with high value for only one interest, such as the breeding site of an endangered animal.

The simplest way to achieve this analysis is by map overlay. First, a base map is prepared to an appropriate scale. Individual data elements (turtle beaches, seabird islands, fishery activities, coral reefs, pinniped rookeries and so on) are then mapped on transparent overlays of this base. An example of this approach which has been taken for the coast of Belize is shown in Figure 2.1 (Gibson, Price and Young, 1993). Next, the transparencies are overlaid to show the areas with resource concentrations. These, together with any sites with specific interests, can then be identified as candidates for selection.

Statistical and issue analyses can also be undertaken. The latter is helpful in the identification and understanding of management issues that are not revealed by statistical analysis but which have significant implications for protected area management (Salm and Dobbin, 1987).

2.2.3 Data synthesis

In this step, maps (graded if possible to high, medium and low levels) of activities, threats to resources, or conflicts between resource protection and developments can be combined with the composite map resulting from the preceding analysis to refine the identification process. This synthesis helps to impose a measure of priority for protection on the candidate sites. For example, the level of dependence by people on a fishery could be used to indicate priority among candidate sites to safeguard fishery resources. Similarly, the vulnerability to some form of threat posed by an activity or a proposed development may determine the choice between two similar candidate sites.

Another application of data synthesis is to develop an understanding of spatial relationships among biological factors (e.g. species), ecological processes (e.g. nutrient transport) and human activities. Overlays of ocean currents, nutrient sources (such as estuaries) and resource distributions can be combined to show the specific support systems for species or ecosystems – all of which are candidates for protection. Also, by combining overlays of human activities, and particularly proposed developments, a picture of potential conflicts between resource management and developments and of priority areas for action can be built up (Ray *et al.*, 1978).

The overlay mapping technique has limitations, and it is important to apply common sense to the interpretation of results. For example, candidate

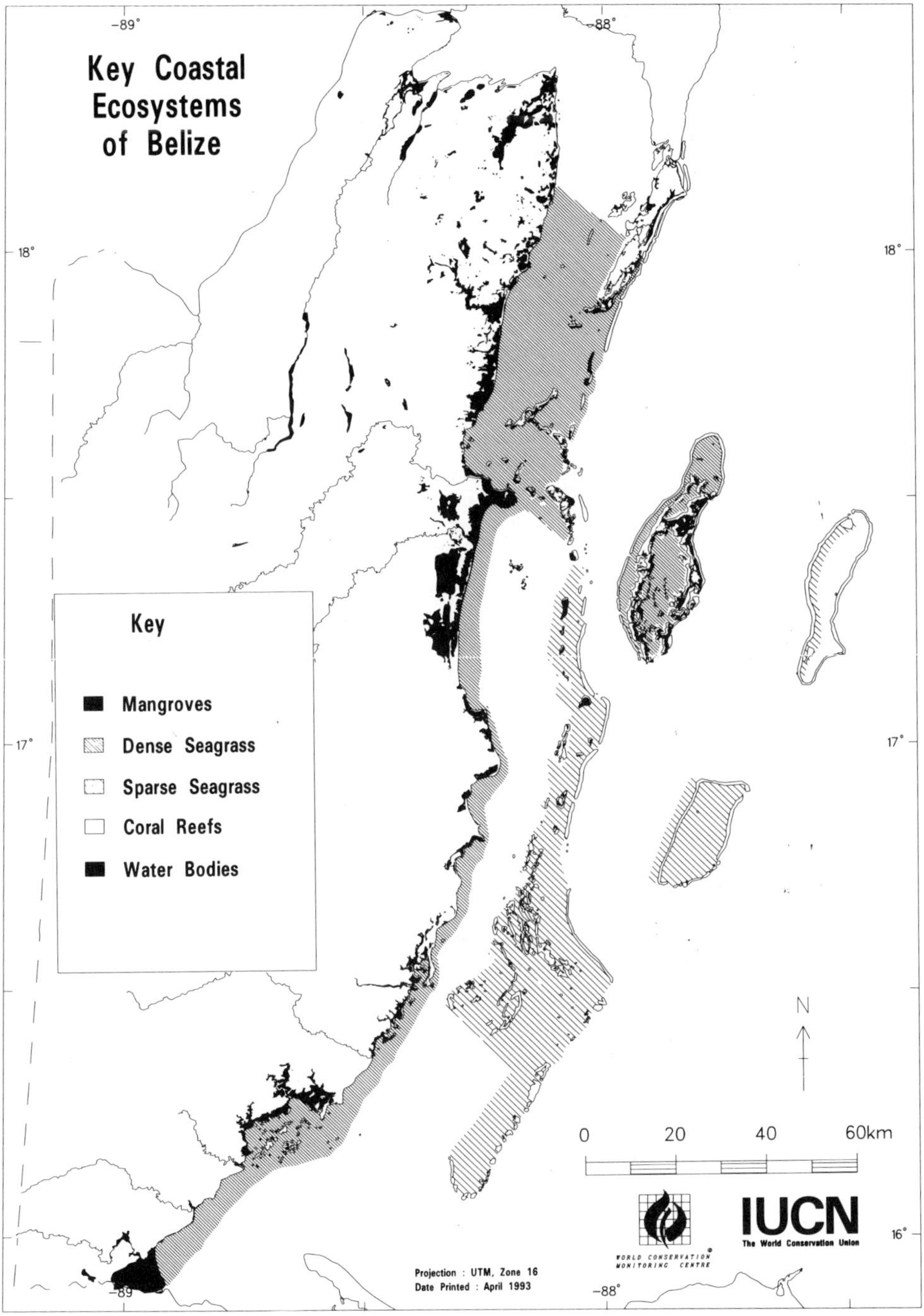

Figure 2.1 Key coastal ecosystems of Belize. See also first page of colour plates between pages 116/117.

sites for protection will tend to be restricted to those without significant levels of human activity. As a result, there will be a tendency to overlook the great potential that some exploited sites may have for restoration. It is also influenced by the amount of data available for particular areas.

As another example, the overlay technique can mislead planners into excluding a site because of a perceived conflict between protection and powerful commercial interests. For example, industrial fisheries in a critical spawning ground may seem like too big a problem to tackle, but protection of stocks during the critical spawning season and allowing fishing during the remainder of the year can be mutually reinforcing activities.

It can be very revealing to compare the candidate sites resulting from data synthesis with the range of established MPAs. As well as indicating the needs for additional MPAs, this comparison may show that some sites do not correspond to critical areas at all.

The identification procedure outline in the data analysis and synthesis steps above will provide a long list of candidate sites for protection. Selection of the specific sites for protection from this list will require the application of criteria.

2.2.4 Application of selection criteria

The first areas to be selected for protection are often so obvious that there is no need to apply criteria except in selecting an appropriate management category (Salm and Clark, 1984). However, the identification and selection of other candidate areas requires the application of criteria.

The application of selection criteria helps to ensure objectivity in the choice of sites for protection. Criteria for identifying and selecting coastal and marine protected areas depend principally on the overall objectives of the programme. If, for example, the objectives are mainly to safeguard areas for tourism and recreation in wilderness settings, the criteria could emphasize accessibility, safety factors (such as the absence of currents and large waves), scenic value, the presence of such other interests as cultural or archaeological sites, and carrying capacity (i.e. the number of visitors the area can sustain without degrading the environment or destroying the quality of the wilderness experience by crowding).

If economic goals (such as coastal protection, maintenance of fisheries and development of tourism and related industries) are the main interest, the criteria might emphasize intensity of resource exploitation, the present and potential economic value of resource and the degree of threat to them. Criteria stressing naturalness, uniqueness and habitat or species diversity are likely to be applied to select sites when the primary objectives are maintenance of genetic diversity, safeguarding of ecological processes and species replenishment.

Criteria have two functions. Initially they serve to assess the eligibility of sites for protected area status. Their principal role, however, is to order eligible sites according to priority in the selection process. The number of sites finally selected for protection will be determined by such factors as national policy, urgency for action, availability of financial and manpower resources and, in the case of some developing countries, the extent of international concern and assistance.

For each site under consideration, the different criteria can be quantified or scored. In the simplest application, a set of selected criteria is scored (e.g. on a scale of 1 to 5 for the lowest to highest value) for each site. The scores are summed for each site and compared, and priority areas are identified on the basis of the highest scores. However, powerful statistical techniques (e.g. principal components analysis, multidimensional scaling or cluster analysis) can be applied to portray graphically the relative similarities and differences between the sites, based on the selected criteria. Different types of criteria (e.g. economic, social, ecological and so on) can be analysed either separately or jointly for a more integrated picture so that sites are arranged into groups. Members of a group share the greatest number of similarities, and differ from members of other groups. Different groups can then be assigned a priority rating taking into account the geographical spread of the candidate sites and the urgency or schedule for establishment. Representative sites from some or all of the groups can then be selected for protected area status.

The next section illustrates some of the applications of criteria and is intended to facilitate the task of defining appropriate criteria. Every effort should be made to limit the number of criteria and make them practical, as well as targeted, so that the selection process meets the specific objectives of the MPA programme.

2.3 CRITERIA FOR THE SELECTION OF MARINE PROTECTED AREAS

The following examples of criteria for the selection of MPAs are based on Salm and Clark (1984) with minor modifications. The section headings (2.3.1–2.3.5) represent programme goals and the numbered headings within each section are the criteria.

2.3.1 Social criteria

Social benefits are measured in the following terms:

1. *Social acceptance* – the degree to which the support of local people is assured. Every effort should be made to canvass local support. When an

area is already protected by local tradition or practice, it should be encouraged, and the area should receive a higher rating. An 'official' protected area designation may not be necessary if local support is high.

2. *Public health* – the degree to which the creation of a MPA may serve to diminish pollution or other disease agents that contribute to public health problems. Granting protected status to contaminated areas, such as shellfish beds or bathing beaches, may result in reduced pollution as the polluting source is identified and controlled as part of the plan for site management.

3. *Recreation* – the degree to which the area is, or could be, used for recreation by country residents. Sites that provide the local community the opportunity to use, enjoy and learn about their local natural environment should rate highly for this criterion.

4. *Culture* – the religious, historic, artistic or other cultural value of the site. Natural areas that also contain important cultural features should be given high ratings as their protection may help to maintain the integrity of the adjacent ecosystem.

5. *Aesthetics* – a seascape, landscape, or other area of exceptional scenic beauty. Natural areas that also contain features of natural beauty should be given higher ratings since the safeguarding of such features often requires that the integrity of adjacent coastal and marine systems be maintained. However, where species diversity and the biological conservation value are low, such areas retain a high value for recreation and tourism.

6. *Conflicts of interest* – the degree to which area protection would affect the activities of local residents. If the area is to be used for recreation purposes, for example, the site should not be a major fishing area and should have few dependent fishermen. In some instances, careful zoning can minimize such conflicts.

7. *Safety* – the degree of danger to people from strong currents, surf, submerged obstacles, waves and other hazards. The principal users will often be swimmers, snorkellers, divers and boaters. It is important that they are able to pursue their activities safely.

8. *Accessibility* – the ease of access across both land and sea. Areas to be used by visitors, students, researchers and fishermen must be accessible to them. The more accessible, the greater the value; but the greater the level of use, the greater the likelihood of conflicting interests and the greater the impact of users. Accessibility weighs high for MPAs with predominantly

social objectives, fairly high for those with economic goals and low for those meeting ecological criteria (section 2.3.2).

9. *Research and education* – the degree to which an area represents various ecological characteristics and can serve for research and demonstration of scientific methods. Areas that clearly demonstrate different habitat types and ecological relationships and are sufficiently large both to serve conservation and to accommodate teaching (i.e. field trips or on-site learning centres) should receive a higher rating.

10. *Public awareness* – the degree to which monitoring, research, education or training within the area can contribute knowledge and appreciation of environmental values and conservation objectives. Areas that can combine such activities as pollution monitoring and education should receive a higher rating.

11. *Conflict and compatibility* – the degree to which an area may help to resolve conflicts between natural resource values and human activities, or the degree to which compatibilities between them may be enhanced. If an area can be used to exemplify the resolution of conflicts in the region, it should receive a higher rating.

12. *Benchmark* – the degree to which the area may serve as a 'control site' for scientific research (i.e. an unmanipulated area used to measure changes elsewhere). Benchmark areas are essential components of an ecological monitoring programme and should receive a higher rating.

2.3.2 Economic criteria

Economic benefits are measured in the following terms:

1. *Importance to species* – the degree to which certain commercially important species depend on the area. Reefs, estuaries or wetlands, for example, may be critical habitats for certain species that breed, rest, shelter or feed there, and that form the basis of local fisheries in adjacent areas. Such habitats need management to support the exploited stocks.

2. *Importance to fisheries* – the number of dependent fishermen and the size of the fishery yield. The greater the dependence of fishermen on an area and the greater its yield of fish, the more important it becomes to manage the area correctly and to ensure sustainable harvest.

3. *Nature of threats* – the extent to which changes in use patterns threaten the overall value to people. Habitats may be threatened directly by destruc-

tive practices, such as fishing with explosives and certain bottom trawls, or by overexploitation of resources. Areas traditionally harvested by local fishermen become important to manage. The numbers of fishermen on these grounds may increase, bringing extra pressure to bear on stocks and habitats. Even if the numbers do not change, the traditional capture methods may be replaced by others that yield more per unit effort (an extreme example is the use of explosives). The stocks of some species may not be capable of withstanding such increased drains on their breeding population. In this way whole species have disappeared from fishing grounds or have become exceedingly rare.

4. *Economic benefits* – the degree to which protection will affect the local economy in the long term. Initially, some protected areas may have a short-lived, disruptive economic effect. Those that have obvious positive effects should have higher ratings (for example, for protecting feeding areas of commercial fish or areas of recreational value).

5. *Tourism* – the existing of potential value of the area to tourism development. Areas that lend themselves to forms of tourism compatible with the aims of conservation should receive a higher rating.

2.3.3 Ecological criteria

The values of ecosystems and their species are measured in the following terms:

1. *Diversity* – the variety or richness of ecosystems, habitats, communities and species. Areas having the greatest variety should receive higher ratings. However, this criterion may not apply to simplified ecosystems, such as some pioneer or climax communities, or areas subject to disruptive forces, such as shores exposed to high-energy wave action.

2. *Naturalness* – the lack of disturbance or degradation. Degraded systems will have little value for fisheries or tourism and will make little biological contribution. A high degree of naturalness scores highly. If restoring degraded habitats is a priority, a high degree of degradation may score highly.

3. *Dependency* – the degree to which a species depends on an area, or the degree to which an ecosystem depends on ecological processes occurring in the area. If an area is critical to more than one species or process, or to a valuable species or ecosystem, it should have a higher rating.

4. *Representativeness* – the degree to which an area represents a habitat type, ecological process, biological community, physiographical feature or

other natural characteristic. If a habitat of a particular type has not been protected, it should have a high rating. (A classification scheme for coastal and marine areas is desirable in applying this criterion.)

5. *Uniqueness* – whether an area is 'one of a kind'. Habitats of endangered species occurring only in one area are an example. The interest in uniqueness may extend beyond country borders, assuming regional or international significance. To keep visitor impact low, tourism may be prohibited but limited research and education permitted. Unique sites should always have a high rating.

6. *Integrity* – the degree to which the area is a functional unit – an effective, self-sustaining ecological entity. The more ecologically self-contained the area, the more likely it is that its values can be effectively protected, and so a higher rating should be given to such areas.

7. *Productivity* – the degree to which productive processes within the area contribute benefits to species or to humans. Productive areas that contribute most to ecosystem sustainment should receive a high rating. Exceptions are eutrophic areas where high productivity may have a deleterious effect.

8. *Vulnerability* – the areas's susceptibility to degradation by natural events or the activities of people. Biotic communities associated with coastal habitats may have a low tolerance to changes in environmental conditions or they may exist close to the limits of their tolerance (defined by water temperature, salinity, turbidity or depth). They may suffer such natural stresses as storms or prolonged emersion that determine the extent of their development. Additional stress (such as domestic or industrial pollution, excessive reductions in salinity and increases in turbidity from watershed mismanagement) may determine whether there is total, partial or no recovery from natural stress, or the area is totally destroyed.

2.3.4 Regional criteria

The contribution of an area to a regional network of protected areas can be assessed in the following terms:

1. *Regional significance* – the degree to which the area represents a characteristic of the region, whether a natural feature, an ecological process or a cultural site. The role the area plays in contributing nutrients, materials or support for species (especially migratory ones) to the region as a whole should be evaluated. Both ecological processes and natural resources are

often shared among nations, so areas contributing to the maintenance of species or ecosystems beyond national boundaries should have higher ratings.

2. *Subregional significance* – the degree to which an area fills a gap in the network of protected areas from the subregional perspective. This contribution may be assessed by comparing the distribution of protected areas with subregional characteristics. If a type of area is preserved in one subregion, that type should also be protected in another subregion.

2.3.5 Pragmatic criteria

The feasibility and appropriate timing of protection can be measured in terms of the following:

1. *Urgency* – the degree to which immediate action must be taken, lest values within the area be transformed or lost. Lack of urgency should not necessarily be given a lower rating since it is often best, and least costly, to protect well in advance of the threat.

2. *Size* – which and how much of the various habitats need to be included in the protected area. Size is an important factor in designing protected areas. It has often been overlooked in the design process, resulting in severe degradation, even total destruction, of protected areas. To receive a high rating, the protected area must be large enough to function as an ecological unit.

3. *Degree of threat* – present and potential threats from direct exploitation and development projects. The further the protected area is from potential sources of accidental poisoning (such as large ports, petroleum deposits or river mouths), the better are the survival prospects of species and communities. However, if an important habitat is severely threatened, it may be important to implement a management plan to reduce the threats to tolerable levels.

4. *Effectiveness* – the feasibility of implementing a management programme. A site that satisfies many criteria, but cannot be adequately managed (i.e. monitored, patrolled and defended) is not of much use. Higher ratings should go to sites that are manageable.

5. *Opportunism* – the degree to which existing conditions or actions already under way may justify further action. An extension of an established protected area should have a higher rating.

6. *Availability* – the degree to which the area is available for acquisition or can be managed satisfactorily by agreement with the owners. The problem of tenure rarely applies to the sea. Beaches also often belong to the central or provincial government. Thus, acquisition of aquatic areas, wetlands and seashores may not be necessary. However, adjacent lands and islands may be privately owned or leased. Generally, to secure long-term control over these areas, the title or lease will need to be bought from current owners. Higher ratings should go to areas owned by state or national governments.

7. *Restorability* – the degree to which the area may be returned to its former natural state. Areas that can increase in productivity or value to important species and processes should receive higher ratings.

2.4 TWO CASE STUDIES ON SITE SELECTION

The approach and selection procedure is described for two nations of the Arabian Peninsula: the Sultanate of Oman and the Kingdom of Saudi Arabia. The approach taken in these two cases has been fairly well documented and may be of value as prototypes for the selection of marine and coastal protected areas elsewhere.

2.4.1 Sultanate of Oman

In 1986 IUCN completed a study defining a system of protected areas for the Sultanate (IUCN, 1986b). This was the first attempt at formulating a coordinated national nature conservation programme which defined conservation policy, legislation, staffing and administration needs, and which identified potential conservation areas. It identified 63 proposed protected areas which include a coastal component, five of which already had protected status. A wider IUCN coastal zone management project carefully studied each proposed protected area in the coastal sector for which a management plan was being prepared. Where necessary, revision to the proposals and zoning details were recommended in the coastal zone management plans (IUCN, 1986a, 1988, 1989, 1990). In addition, management plans were prepared for three of the decreed protected areas (Salm 1986, 1989).

The Sultanate's five existing protected areas covered a range of environments, species, objectives and management issues. The Daymaniyat Islands are a remote group of small islands with minimal conflicts and management issues. They are of global importance for the highly endangered hawksbill turtle population that nests there, have well developed coral reefs, thousands of nesting seabirds, and nesting ospreys and sooty falcons (Salm, 1986).

In contrast, the Q'urm mangroves lie in the heart of the prime residential area and the adjacent seashore is heavily used for recreational activities.

Considerable care has been taken to avoid endangering the mangrove environment during the course of urban development, particularly in the building of the highway immediately inland of the estuary. Khawr Salalah is an important estuary for migrant waterfowl and waders. It too is hemmed in by development, but is fenced and guarded against intrusion.

The remaining two protected areas, Ra's al Hadd and Ra's al Junayz, have beaches with one of the largest populations of nesting green turtles in the Indian Ocean. These mainland sites are intensively used for both recreation and fisheries but the area was closed to campers because of excessive disturbance to nesting turtles. This is an area of maximum conflicts between conservation objectives and escalating use. A conservation management plan for the two protected areas was formulated (Salm, 1989) and is being implemented.

Each of the five sites has its intrinsic values and would score highly on a combined list of ecological, social, regional and economic criteria. However, urgency is the one criterion that determined their priority for designation as protected areas. Each site was under severe threat from development and related activities.

The government's commitment to coastal zone management has resulted in a significant contribution to conservation of coastal and marine environments. Coastal zone management is a multi-pronged approach which, through a combination of policy, law, protected areas and issue-specific actions, manages the entire coastal zone as a vast multiple use reserve (Salm, 1988).

2.4.2 Kingdom of Saudi Arabia

As part of its long-term planning, Saudi Arabia's Meteorology and Environmental Protection Administration (MEPA) commissioned an appraisal of natural resources and management requirements for both the Arabian Gulf and Red Sea coasts (Ormond *et al.*, 1984a,b,c; IUCN/MEPA 1987a,b,c; Price *et al.*, 1987; Sheppard *et al.*, 1992; Price, 1993;). The studies, in conjunction with earlier initiatives of MEPA, the Environmental Protection Coordination Committee (EPCCOM) and IUCN, identified a total of 11 environmentally sensitive areas (ESAs) plus seven recreational areas in the Gulf. A further 46 ESAs that include some recreation areas were identified in the Red Sea. ESAs are areas that merit special protection or management and are nearly synonymous with critical habitats (Al Gain *et al.*, 1987). These represent potential sites for a system of protected areas, known locally as marine protectorates.

ESAs in the Red Sea were identified primarily on the basis of the unique and valuable resources or important ecological processes concentrated in them. On the Saudi Arabian Gulf coast, where coastal development has been much greater, the selection procedure for ESAs was taken a step further. Information on the location of key natural resources was compared with the location of coastal and marine uses and activities. This identified the main areas of resource/use conflict, and hence where site specific management measures were needed most urgently. Other factors were also taken into account (IUCN/MEPA, 1987a).

This proposed protected area system is more conceptual than specific. Further analysis of the main socio-economic issues is needed, together with more precise formulation of management objectives. These ESAs will then need to be built into the framework of an overall national coastal zone management programme. Until this happens it was recommended that the ESAs receive temporary protected status. The 57 ESAs also need to be studied to determine how they might be combined into larger areas to facilitate management.

Since then the National Commission for Wildlife Conservation and Development (NCWCD) has evaluated marine and terrestrial areas for inclusion in the National Protected Areas System Plan (Child and Grainger, 1990). These include the Farasan Islands, declared a protected area in 1989. Management plans are being developed for several areas. In 1994 there were three officially declared protected areas in Saudi Arabia which are either marine or terrestrial, but extend to the coast. All are in the Red Sea region: the Asir National Park, the Farasan Islands Protected Area and the Yanbu Royal Commission Conservation Area (Chifings, 1994; Halim *et al.*, in press) Following the 1991 Gulf War the NCWCD, in collaboration with the European Commission, has developed plans for establishing a Marine Habitat and Wildlife Sanctuary for Dawhat ad-Dafi, Dawhat al-Musallamiya and offshore islands (Krupp and Jones, 1993; Feltkamp and Krupp, 1994). Problems associated with the assessment of the Gulf environment and the need for new, extended forms of governance are addressed by McGlade and Price (1993).

REFERENCES

Al Gain, A., Clark, J. and Chiffings, T. (1987) *A coastal management programme for the Saudi Arabian Red Sea coast*, Coastal Zone '87, Proceedings of the Fifth Symposium on Coastal Zone Management, Amer. Soc. Civ. Eng., New York, 1673–1680.

Chiffings, A.W. (1994) *The global representative system of marine protected areas; North Western Indian Ocean (region 8b)*. Preliminary working document, CNPPA/World Bank Biodiversity Project.

Child, G. and Grainger, J. (1990) *A system plan for protected areas for wildlife and sustainable rural development in Saudi Arabia*, IUCN/NCWCD, 335 pp.

Feltkamp, E. and Krupp, F. (eds) (1994) *Establishment of a Marine Habitat and Wildlife Sanctuary for the Gulf Region. Final Report for Phase II*, CEC/NCWCD (Saudi Arabia) report, 675 pp.

Gibson, J.P., Price, A.R.G. and Young, E. (1993) *Guidelines for Developing a Coastal Zone Management Plan for Belize. The GIS Database*. A Marine Conservation and Development Report, IUCN, Gland, Switzerland, vi +11 pp., 9 maps.

Halim, Y., Awad, H.E., El-Sayed, M. Kh. *et al.* (in press) *The state of the marine environment in the Red Sea and Gulf of Aden*, UNEP Regional Seas Report.

IUCN (1986a) *Oman coastal zone management plan: Greater capital area*, IUCN, Gland, Switzerland, 78 pp.

IUCN (1986b) *Proposals for a system of nature conservation areas*. IUCN, Gland, Switzerland, 469 pp.

IUCN/MEPA (1987a) *Saudi Arabia: An assessment of biotopes and management requirements for the Arabian Gulf*, MEPA Coastal and Marine Management Series, Report No. 5, 248 pp.

IUCN/MEPA (1987b) *Saudi Arabia: An assessment of coastal zone management requirements for the Red Sea*, MEPA Coastal and Marine Management Series, Report No. 3, 105 pp.

IUCN/MEPA (1987c) *Saudi Arabia: An assessment of national coastal zone management requirements*, MEPA Coastal and Marine Management Series, Report No. 7, 41 pp.

IUCN (1988) *Oman coastal zone management plan: Quriyat to Ra's al Hadd*, IUCN, Gland, Switzerland, 57 pp.

IUCN (1989) *Oman coastal zone management plan: Dhofar*, Vol. 1, Action Plan, IUCN, Gland, Switzerland, 37 pp.

IUCN (1990) *Oman coastal zone management plan: Musandam*, IUCN, Gland, Switzerland.

Kelleher, G. and Kenchington, R. (1992) *Guidelines for Establishing Marine Protected Areas*, A Marine Conservation and Development Report, IUCN, Gland, Switzerland, vii + 79 pp.

Krupp, F. and Jones, D.A. (1993) The creation of a marine sanctuary after the Gulf War oil spill. *Mar. Poll. Bull.* **27**, 315–323.

McGlade, J.M. and Price, A.R.G. (1993) Multi-disciplinary modelling: an overview and practical implications for the governance of the Gulf region. *Mar. Poll. Bull.*, **27**, 361–365.

Ormond, R.F.G., Dawson-Shepherd, A.R., Price, A.R.G. and Pitts, R.J. (1984a) *Report on the distribution of habitats and species in the Saudi Arabian Red Sea*, Parts 1 and 2, IUCN, Gland, Switzerland, 247 pp.

Ormond, R.F.G., Dawson-Shepherd, A.R., Price, A.R.G. and Pitts, R.J. (1984b) *Distribution of habitats and species along the southern Red Sea coast of Saudi Arabia*, IUCN, Gland, Switzerland.

Ormond, R.F.G., Dawson-Shepherd, A.R., Price, A.R.G. and Pitts, R.J. (1984c) *Management of Red Sea coastal resources: recommendations for protected areas*, IUCN, Gland, Switzerland, 113 pp.

Price, A.R.G. (1990) Rapid assessment of coastal zone management requirements: Case study in the Arabian Gulf. *J. Ocean Shorel. Manage*, **13**(1990), 1–19.

Price, A.R.G. (1993) The Gulf: Human impacts and management initiatives. *Mar. Poll. Bull.*, **27**, 17–27.

Price, A.R.G., Chiffings, T.W., Atkinson, T.J. and Wrathall, T.J. (1987) *Appraisal of resources in the Saudi Arabian Gulf*, Coastal Zone '87, Proceedings of the Fifth Symposium on Coastal Zone Management, Amer. Soc. Civ. Eng., New York, 1031–1045.

Ray, G.C. (1976) Critical marine habitats, in *Proceedings of an International Conference on Marine Parks and Reserves*, IUCN Publication (n.s.) No. 37, Gland, Switzerland, pp. 15–64.

Ray, G.C., Dobbin, J.A. and Salm, R.V. (1978) Strategies for protecting marine mammal habitats. *Oceanus*, **21**(3), 55–67. (Reprinted 1981 in *Prioroda*, **8**, 95–101.)

Salm, R.V. (1986) *The proposed Daymaniyat Islands National Nature Reserve management plan*, IUCN Coastal Zone Management Project, Oman, 45 pp.

Salm, R.V. (1987) *Coral reef management planning: Indonesia and Oman*. Paper presented at the Workshop on Innovative Ways of Planning and Managing the Great Barrier Reef Marine Park, 20–22 July, 1988, Townsville, Australia, 6 pp.

Salm, R.V. (1988) Coral reef management planning: Indonesia and Oman. Paper presented at the Workshop on Innovative ways of Planning and Managing the Great Barrier Reef Marine Park, 20–22 July, 1988, Townsville, Australia, 6 pp.

Salm, R.V. (1989) A proposed management plan for the turtle nesting beaches in the Ra's al Hadd National Scenic Reserve and Ra's al Junayz National Nature Reserve. IUCN Coastal Zone Management Project. *CZMP*, **3**, I 11, 49 pp.

Salm, R.V. and Clark, J.R. (1984) *Marine and Coastal Protected Areas: A Guide for Planners and Managers*, IUCN, Gland, Switzerland, xiii + 302 pp.

Salm, R.V. and Dobbin, J.A. (1987) *A coastal zone management strategy for the Sultanate of Oman*, Coastal Zone '87, Proceedings of the Fifth Symposium on Coastal Zone Management, Amer. Soc. Civ. Eng., New York, **1**, 97–106.

Salm, R.V. and Dobbin, J.A. (1989) *Coastal zone management, planning and implementation the Sultanate of Oman*, Coastal Zone '89, Proceedings of Sixth Symposium on Coastal and Ocean management, Amer. Soc. Civ. Eng., New York, 72–78.

Salm, R.V. and Dobbin, J.A. (1993) Planning, management and administrative processes for marine protected areas, in *Application of the biosphere reserve concept to coastal marine areas* (eds A.R.G. Price and S.L. Humphrey). Papers presented at the UNESCO/IUCN San Francisco Workshop, August 1989. A Marine Conservation and Development Report, IUCN, Gland, Switzerland, pp. 57–65.

Sheppard, C.R.C., Price, A.R.G. and Roberts, C.M. (1992) *Marine Ecology of the Arabian Region. Patterns and Processes in Extreme Tropical Environments*, Academic Press, London, 359 pp.

3

Legislative requirements

John Gibson and Lynda Warren

The legislative requirements for the creation of marine protected areas depend on a combination of international and national law. The first part of this chapter discusses the legal framework within which such areas must operate: it examines the general principles of international law with which they must comply, reviews the global and regional conventions providing for them, and considers the role of national law in their establishment. The second part of the chapter is devoted to a case study of the legal implementation of marine nature reserves in the UK.

Nautical miles

The term 'nautical mile' in the following sections is often used in international treaties covering areas of sea. Specifically it represents one minute of latitude measured at the equator. For scientific purposes (other than navigation), one nautical mile is defined as being exactly 1852 m (6076 ft; 1.1508 statute miles).

3.1 THE LEGAL FRAMEWORK

3.1.1 General principles of the international law of the sea

The principal rules of customary international law governing the uses of the sea were codified at Geneva in 1958. They were subsequently reviewed and extended by the Third United Nations Conference on the Law of the Sea, which produced the comprehensive UN Convention on the Law of the Sea (cited as LOSC), which was agreed in 1982. This Convention was signed by 159 States and other entities, but did not become legally binding until a year after it had been ratified by 60 States. The 60th instrument of ratification was deposited on 16 November 1993, and the Convention entered into force on 16 November 1994.

The extent of the control that a State can assert in a marine protected

Marine Protected Areas: principles and techniques for management. Edited by Susan Gubbay.
Published in 1995 by Chapman & Hall, London. ISBN 0 412 59450 1.

area depends on the location of the area to be protected. States can, of course, enact legislation that will bind their own subjects and vessels anywhere in the world, but discriminatory measures based on nationality would be a partial and unsatisfactory method of marine conservation. International law recognizes different types of maritime zone that may be subject to varying degrees of coastal jurisdiction; of these, the most important for the purposes of marine protected areas are internal waters, the territorial sea and the exclusive economic zone. Beyond these zones lie the high seas, which are open to all States and incapable of national sovereignty (LOSC, articles 86–87, 89); the freedom of the high seas, which includes rights of navigation and fishing, precludes the kind of interference with foreign vessels that would be necessary for marine protected areas. Specific international agreements, which could only be binding on the contracting parties, would be necessary for the creation of such areas in the high seas.

Internal waters

Internal waters are all those on the landward side of the baseline from which the breadth of the territorial sea is measured (LOSC, art. 8); the position of those baselines is explained later. Within its own internal waters, a State enjoys absolute sovereignty, except in so far as it may have undertaken treaty obligations to admit foreign vessels to its ports. Thus, the designation of marine protected areas in internal waters is essentially a matter for national legislation, the requirements of which are discussed in section 3.2.

Territorial sea

The sovereignty of a State also extends to its territorial sea (LOSC, art. 2), but this differs from internal waters in that it is subject to a right of 'innocent passage' by foreign vessels (LOSC, art. 17). The normal baseline of the territorial sea is the low-water line along the coast as marked on large-scale charts officially recognized by the coastal State, although it is left to the State itself to select the tidal criteria (LOSC, art. 5). However, because a low water baseline can produce impractical results with some coastal configurations, alternative baselines are allowed in special circumstances; these often have the effect of enclosing larger areas as internal waters and extending the scope of national sovereignty. Thus, where a coastline is deeply indented or fringed with islands, straight baselines following its general direction may be employed instead (LOSC, art. 7); over 50 States currently use such a system for at least part of their coast. Elsewhere, straight lines up to 24 nautical miles long may be drawn across well-marked bays belonging to a single State, if the area of the bay is at least as large as that of a semicircle whose diameter is a line across the mouth (LOSC, art. 10); if these lines enclose areas that were not previously internal waters, a right of innocent passage continues to exist there (LOSC, art. 8). There are also some 'historic bays', which are claimed in their entirety by States under customary international law on the basis of the long assertion and acceptance of national authority.

Rivers flowing directly into the sea are also closed by a straight baseline across their mouth (LOSC, art. 9) and the outermost permanent harbour works forming an integral part of a harbour system are regarded as part of the coast (LOSC, art. 11). The Law of the Sea Convention requires the publication of charts or lists of geographical coordinates identifying straight baseline systems and closing lines across river mouths and bays (LOSC, art. 16).

The delimitation of the territorial sea is further complicated by the existence of islands, reefs and low-tide elevations. Naturally formed islands that are above water at high tide, irrespective of size, possess their own territorial sea (LOSC, art. 121); where islands are situated on atolls or have fringing reefs, the baseline is the seaward low-water line of the reef (LOSC, art. 6). Moreover, low-tide elevations, which are naturally formed areas of land surrounded by water that are exposed at low tide but submerged at high tide, may also be used as baselines if they are situated within a distance not exceeding the breadth of the territorial sea (LOSC, art. 13). The Law of the Sea Convention permits States to establish a limit up to 12 nautical miles from the baselines (LOSC, art. 3). In the absence of mutual agreement, States with opposite or adjacent coasts may not generally extend their territorial sea beyond the median line (LOSC, art. 15). 116 States have now declared a 12-mile (19 km) territorial sea, but 11 still retain a narrower limit, while 16 make more extensive claims between 20 and 200 miles (32 and 322 km) (Law of the Sea Bulletin, 1993).

The right of innocent passage in the territorial sea entitles foreign ships to navigate through those waters. Such passage must be continuous and expeditious, but it may include stopping and anchoring in so far as these are incidental to ordinary navigation or are rendered necessary by force majeure or distress (LOSC, art. 18). Passage is deemed innocent as long as it is not prejudicial to the peace, good order or security of the coastal State. The Law of the Sea Convention contains a list of non-innocent activities, including fishing and acts of wilful and serious pollution (LOSC, art. 19). A coastal State may take whatever steps are necessary to prevent passage that is not innocent (LOSC, art. 25), and can thus exclude such vessels from the territorial sea. Moreover, foreign ships exercising the right of innocent passage must comply with legislation enacted by the coastal State in conformity with international law, and the Law of the Sea Convention expressly permits such legislation for the conservation of the living resources of the sea, the preservation of the environment and the prevention of pollution (LOSC, art. 21). Accordingly, although a State cannot prevent genuinely innocent passage by foreign ships through its territorial sea, it can regulate the manner in which the right is exercised. This should permit, for example, the imposition of speed limits in a marine protected area, but it is less certain whether such a ship can be excluded altogether. A coastal State is forbidden to hamper innocent passage (LOSC, art. 24), but an exclusion zone would not necessarily do so if convenient alternative routes are available.

Straits

A lesser degree of control is permissible in straits used for international navigation between parts of the high seas or exclusive economic zones, such as the Straits of Gibraltar and Dover. In these, a new right of 'transit passage' is introduced by the Law of the Sea Convention, conferring freedom of navigation and overflight on foreign ships and aircraft for the purpose of continuous and expeditious transit (LOSC, art. 38). There is no requirement that passage should be innocent, but ships must refrain from activities not incidental to normal transit, and must comply with international regulations, procedures and practices for safety at sea and the control of pollution (LOSC, art. 39). However, the scope of national legislation is limited, and the only available measures that would be useful in marine protected areas are the prevention of fishing and the application of international standards on navigational safety and pollution (LOSC, art. 42).

Archipelagos

Another innovation in the Law of the Sea Convention is a special regime for archipelagic States, such as Fiji, Indonesia, Mauritius and the Philippines. An archipelago is defined as a group of islands, interconnecting waters and other natural features which are so closely interrelated that they form an intrinsic geographical, economic and political entity (LOSC, art. 46). Archipelagic States are permitted to draw straight baselines joining the extremities of their outermost islands and drying reefs, provided that they include the main islands and enclose an area in which the ratio of water to land is between 1:1 and 9:1; such baselines must follow the general configuration of the archipelago, and may not generally exceed 100 nautical miles, although up to 3% may have a maximum length of 125 miles (200 km) (LOSC, art. 47). The breadth of the territorial sea and other maritime zones is measured from these baselines (LOSC, art. 48). However, archipelagic baselines do not replace internal waters, and a State may also draw closing lines across river mouths, bays and harbour works according to the ordinary principles (LOSC, art. 50). Within archipelagic waters, prior agreements with other

States, traditional fishing rights and existing submarine cables must be respected, and foreign ships enjoy a right of innocent passage (LOSC, arts 51–52). In addition, an archipelagic State may designate sea lanes and traffic separation schemes in consultation with the International Maritime Organization (IMO), and these will be subject to a right of 'archipelagic sea lanes passage' equivalent to the right of transit passage through international straits (LOSC, arts 53–54).

Exclusive economic zone

The Law of the Sea Convention approved the concept of the exclusive economic zone (EEZ), which stretches from the outer limit of the territorial sea to a maximum distance of 200 nautical miles from the territorial sea baselines (LOSC, arts 55, 57). A minor qualification is that rocks incapable of sustaining human habitation or economic life of their own cannot possess an EEZ (LOSC, art. 121). The delimitation of an EEZ between States with opposite or adjacent coasts should be decided by agreement or a dispute settlement procedure (LOSC, art. 74). Charts or lists of geographical coordinates must be published showing the outer limits of the zone (LOSC, art. 75), and 89 States have already declared an EEZ, while another 21 have limited their claims to exclusive fishery zones (Law of the Sea Bulletin, 1993).

Within an EEZ, the coastal State has sovereign rights for the purpose of exploring, exploiting, conserving and managing the living and non-living resources of the sea-bed, its subsoil and the superjacent waters (LOSC, art. 56). It also has jurisdiction in various matters; these include the protection and preservation of the marine environment, but only as provided for in other 'relevant provisions' of the Convention, which relate to pollution from sea-bed activities, from dumping and from vessels (LOSC, Part XII). The coastal State is empowered to enforce its associated legislation by whatever measures are necessary (LOSC, art. 73). Other States, however, enjoy freedom of navigation and overflight in the EEZ, together with the right to lay submarine cables and pipelines, although they are required to comply with compatible laws and regulations adopted by the coastal State (LOSC, art. 58).

The combined effect of these provisions is unclear, since conflicts of use are inevitable. Thus, it remains to be decided how far the navigation of foreign ships in a marine protected area within an EEZ can be controlled by national legislation (Burke, 1981). The Law of the Sea Convention merely provides that disputes should be resolved on the basis of equity and in the light of all relevant circumstances, taking into account the relative importance of the interests involved to the parties as well as to the international community (LOSC, art. 59).

It should also be noted that the rights of coastal States in the EEZ may be restricted by treaty. For example, Member States of the European Union have ceded jurisdiction over fisheries to the EU itself, and cannot impose unilateral restrictions on fishing within the 200-mile limit on anyone other than their own fishermen. Controls over fishing in a marine protected area inside the European Union's exclusive fishery zone would need to be made by the EU rather than the Member State, and even legislation confined to the national waters of the 12-mile belt must be notified to the European Commission.

Continental shelf

The continental shelf is defined in the Law of the Sea Convention as the natural prolongation of the land territory to the outer edge of the continental margin or a minimum distance of 200 nautical miles from the territorial sea baselines, subject to a maximum of 350 miles (560 km) from the baselines or 100 miles (160 km) from the 2500 m isobath (LOSC, art. 76). The continental shelf between opposite or adjacent States should be determined by mutual agreement, although in cases of dispute the Law of the Sea Convention specifies a settlement procedure (LOSC, art. 83). The Convention also requires the publication of charts or lists of geographical coordinates identifying the outer limits of the continental shelf (LOSC, art. 84). A coastal State has sovereign rights over the continental shelf for the purpose of exploring it and exploiting its natural resources, but these are confined to mineral and other non-living resources together with sedentary species of living organisms (LOSC, art. 77). Thus, control over a marine protected area on the basis of its status as part of the continental shelf must be limited to the regulation of interference with those particular resources. Otherwise, the Law of the Sea Convention only confers jurisdiction over dumping there (LOSC, art. 210). Moreover, the rights of a coastal State in the continental shelf do not affect the legal character of the superjacent waters or the airspace, and must not infringe navigation or other freedoms of the high seas (LOSC, art. 78). In practice, the continental shelf of many States is now coexistent with an exclusive economic zone, which provides much wider jurisdiction, but the concept of the continental shelf remains significant where no EEZ has been declared or the continental shelf extends beyond such a zone.

Marine conservation in the Law of the Sea Convention

The Law of the Sea Convention deals with conservation in somewhat general terms. It imposes a basic obligation on States to protect and preserve the marine environment, and requires them individually or jointly to take action to prevent, reduce and control pollution from any source

(LOSC, arts 192, 194); such measures specifically include those necessary to protect and preserve rare or fragile ecosystems as well as the habitat of depleted, threatened or endangered species and other forms of marine life. States are also expected to cooperate on a global or regional basis in formulating international standards for the protection and preservation of the marine environment, taking into account characteristic regional features (LOSC, art. 197). There is no direct reference to marine protected areas, but the Law of the Sea Convention does not attempt to dictate specific technical measures, which are more appropriately left to specialist conventions.

3.1.2 Marine protected areas in international law

There are numerous international conventions and other multilateral instruments that seek to protect particular marine areas. Some of these are global in their coverage, while others are of regional application. The effect of these agreements will now be discussed.

Special areas under MARPOL

Although the International Convention for the Prevention of Pollution from Ships 1973 (MARPOL) prescribes standards applicable throughout the world, it also imposes stricter controls in certain 'special areas'; these are areas that are considered most sensitive, generally because they are almost totally enclosed by land and subject to severe pressure from shipping. The special areas currently in force under Annex I of MARPOL, which deals with oil pollution, are the Mediterranean, Baltic, Black Sea and Antarctic. Annex II, which concerns pollution by noxious liquid substances in bulk, identifies two special areas, the Baltic and the Black Sea, while those introduced under Annex V relating to pollution by garbage are the Baltic, North Sea, Antarctic and Wider Caribbean.

Ramsar sites

The Ramsar Convention on Wetlands of International Importance especially as Waterfowl Habitat 1971 (Lyster, 1985) protects areas of marsh, fen, peatland or water, including areas of marine water whose depth does not exceed 6 m at low tide. It requires contracting States to designate at least one such site within their territory for inclusion in a list of internationally important wetlands. The boundaries may incorporate riparian and coastal zones adjacent to the wetlands, together with islands and bodies of marine water deeper than 6 m lying within them. Each State must formulate its planning so as to promote the conservation of those sites and the wise use of wetlands in its territory, and must establish nature reserves on wetlands. By 31

December 1994, 85 States were party to the Ramsar Convention, and had designated 718 wetlands covering more than 43 million hectares. However, Ramsar sites depend for their protection on the effectiveness of national legislation, and the Convention, which applies to inland as well as coastal areas, is necessarily limited both in its seaward extent and in the types of marine habitat that it covers.

Areas to be avoided

Another approach is the designation of 'areas to be avoided', which are established by resolutions of the Assembly of the International Maritime Organization, and have been declared in 26 regions (IMO, 1991). They are areas that should not be entered by any or certain categories of vessel. While some are designated for reasons of navigational safety, others are expressly justified as avoiding the risk of pollution and damage to the environment: these are near the Rochebonne Shelf (France), Cape Terpeniya (Sakhalin, former USSR), the North-west Hawaiian Islands, the Great Barrier Reef, the Bermuda Islands, Grassholme Island (Wales), Aldabra and Mahe Islands (Seychelles), California and Florida Coasts (United States), Fasht Buldani (Saudi Arabia), Three Kings Island (New Zealand) and Voriai Sporadhes (Greece). However, since areas to be avoided are not based on treaty, they only have advisory status and are not legally binding.

Particularly sensitive sea areas

The International Maritime Organization has developed guidelines for the identification of fragile ecosystems, such as mangroves and coral reefs, which may be designated as 'particularly sensitive sea areas' because of their vulnerability to pollution from ships. Only one such area, the Great Barrier Reef in Australia, which was designated in 1991, has so far been declared. Designation does not in itself provide legal protection, which depends on the introduction and enforcement of national or international legislation. However, the status of particularly sensitive sea areas is recognized in Agenda 21, adopted by the United Nations Conference on Environment and Development in 1992. Chapter 17, paragraph 17.30 of Agenda 21 calls on States to assess pollution from ships in particularly sensitive sea areas identified by IMO and to implement applicable measures there to ensure compliance with international regulations.

Protected areas under the UNEP Regional Seas Programme

The most comprehensive legislative measures for marine protected areas have been taken under the Regional Seas Programme of the United

Nations Environment Programme (UNEP). Within the framework of wide-ranging action plans, international conventions have so far been adopted for eight regions: Mediterranean, Persian/Arabian Gulf, West and Central Africa, South-East Pacific, Red Sea and Gulf of Aden, Wider Caribbean, Eastern Africa and South Pacific; conventions have yet to be agreed for the East Asian and South Asian Seas regions. Additional protocols dealing with specially protected areas have already been concluded for the Mediterranean, East African region, South-East Pacific and Wider Caribbean. These impose duties on the contracting parties to establish protected areas and describe the kind of measures to be taken in conformity with the rules of international law. The West and Central African Convention and the South Pacific Convention also contain articles calling for the designation of such areas, but they do not elaborate on the requirement.

There are strong similarities between the protocols, which all provide for the regulation of activities, the creation of buffer zones, international cooperation over frontier protected areas and the publication of information. However, they differ in some respects, and the Wider Caribbean Protocol, which is the latest, contains the most detailed provisions, including a planning and management regime (Freestone, 1990). There are also differences in their seaward extent: the Mediterranean Protocol is confined to the territorial waters of the contracting parties, but the East African Protocol includes all the marine and coastal environment under their jurisdiction; the Wider Caribbean Protocol applies within 200 nautical miles of the coast, whereas the South-East Pacific Protocol covers not only the 200-mile maritime zone, but also the entire continental shelf when the contracting parties extend it beyond that distance. On the landward side, all four protocols include coastal areas and related internal waters, although their precise definitions vary. There are further variations in the degree of control over navigation envisaged by the protocols. The Mediterranean Protocol allows the regulation of passage by ships, including stopping or anchoring; the Wider Caribbean Protocol stresses that this is without prejudice to the rights of innocent passage, transit passage, archipelagic sea lanes passage and freedom of navigation in accordance with international law. In contrast, the Eastern African Protocol provides only for the regulation of pleasure craft activities, while the South-East Pacific Protocol does not expressly specify controls over vessels except in relation to pollution and the transport of hazardous substances. Moreover, while the protocols provide a catalyst for action, and encourage the adoption of consistent standards and a common approach, they do not in themselves confer the legal powers to create marine protected areas, and it is for the participating States to pass the necessary domestic legislation to achieve this.

The North Sea

The Hague Declaration on the Protection of the North Sea, which was signed in 1990, constitutes a common declaration of political will, and is arguably not legally binding on its signatories. It provides targets for the reduction of inputs of hazardous substances and nutrients into the North Sea, the phasing out of PCBs, and the limitation of dumping and incineration. It also contains commitments for the prevention of pollution from ships and offshore installations, the control of radioactive wastes, the improvement of airborne surveillance, the enhancement of scientific knowledge, the extension of coastal state jurisdiction, the salvage of sunken ships and hazardous cargoes, the management of fisheries, and the provision of information and consultation.

In addition, the Declaration includes three provisions of particular relevance to marine protected areas. First, the North Sea States have agreed to give high priority to measures likely to have a special significance for the Wadden Sea, including the designation of important coastal and estuarial wetlands as Ramsar sites and nature reserves. Second, they undertake to improve the protection of marine wildlife, *inter alia* by identifying coastal, estuarial and open sea areas of national or international importance, and applying appropriate conservation measures to them; these may include designation under the Ramsar Convention or the EC Birds and Habitats Directives, mandatory notification of potentially damaging operations in coastal sites and consultation offshore, legal controls over particular sea uses in designated areas, and measures to conserve or restore common seal populations. Third, a memorandum of understanding commits the signatory States to cooperate in order to achieve a favourable conservation status for small cetacean populations in the North Sea, and the management measures proposed include habitat protection by curbing disruptive activities and preventing the release of harmful substances.

Other protected areas

The Helsinki Convention on the Protection of the Marine Environment of the Baltic Sea Area 1992, which replaces a 1974 treaty, contains detailed controls over hazardous substances and over pollution from land-based sources, ships, pleasure craft, dumping and sea-bed activities, and provides for cooperation in combating marine pollution. It also imposes a general duty to conserve natural habitats and biodiversity. The Helsinki Convention relies on the littoral States to implement its provisions in their territorial sea and internal waters through national legislation. Some of its measures are more stringent than corresponding global and regional conventions. Another multilateral Convention on the Protection of the Black Sea Against Pollution was concluded at Bucharest in 1992. There have also been less ambitious bilateral agreements in 1974 between Italy and Yugoslavia concerning pollution of the Adriatic Sea, and between Denmark and Sweden relating to the Sound.

A different approach is adopted by the Canberra Convention on the Conservation of Antarctic Marine Living Resources 1980, which empowers the Commission appointed under that treaty to designate special areas for protection and scientific study as part of the conservation measures for the Antarctic marine ecosystem. Most wildlife conventions, however, do not confer powers, but instead impose duties on national governments (Lyster, 1985). Thus, the Bonn Convention on the Conservation of Migratory Species of Wild Animals 1979 requires the range States of specific endangered species of migratory wild animals, including some marine species, to endeavour to conserve and restore their important habitats; in addition, listed migratory species that have an unfavourable conservation status may be the subject of international agreements, which should also provide for the protection of habitats. Likewise, the Berne Convention on the Conservation of European Wildlife and Natural Habitats 1979 obliges contracting parties to take legislative and administrative measures to conserve the habitats of wild flora and fauna, and to give special attention to the protection of important areas for specified migratory species.

Within the European Union, similar obligations in relation to wild birds are created by the 1979 Birds Directive. Member States must classify the most suitable sites in their land and sea territory as special protection areas for the conservation of listed species of wild birds, paying particular attention to wetlands. In addition, under the 1992 EC Habitats Directive, which applies not only to land but also to maritime areas under the sovereignty or jurisdiction of Member States, special areas of conservation must be established for other species of flora and fauna.

3.1.3 National legislation

This chapter has explained how the general principles of international law provide a framework within which marine protected areas may function, and that specific multilateral agreements may impose duties upon participating States to establish them, but that international law alone does not contain the legislative and enforcement powers to protect such areas. The process of designation and regulation must therefore depend upon national legislation by coastal States. In this context, national legislation is likely to be a two-tier process, involving both primary and secondary measures. Because the requirements of marine protected areas are complex and vary according to their location, primary legislation such as a statute or Act of Parliament is likely to be too cumbersome to contain all the necessary detail; greater flexibility is achieved by the passage of enabling legislation, which determines a common jurisdictional structure, but delegates power to identify and regulate areas to a government minister or public authority. These subordinate powers are exercisable through secondary legislation, which may take the form of a ministerial order or statutory instrument. The choice of delegate is an important political decision, although primary legislation sometimes avoids the issue by using a generic title like 'the Secretary of State'. It is, however, particularly important that enactments conferring statutory powers on officials should be clear and comprehensive, since the doctrine of *ultra vires* prevents them from taking any action that is not expressly or impliedly authorized.

Primary legislation should prescribe the purposes for which marine protected areas may be established and specify their potential geographical scope, but the boundaries of individual sites are normally defined in secondary legislation. National practice differs according to whether there is a single legislative regime for both marine and terrestrial protected areas, as in Australia, or they are covered by separate legislation, as happens in the UK. The integrated approach seems preferable, because it recognizes the interdependent nature of coastal zones. Separate systems make it difficult to control activities occurring either on land or sea that have an adverse impact on the other medium, since it may prove legally impossible to use the legislation for one regime to protect an area outside its own jurisdiction. Moreover, legal boundaries on the foreshore tend to be based on average tidal criteria that are primarily used for navigation or the demarcation of property and do not correspond to the location of marine ecosystems.

The power to regulate marine protected areas must be provided by the primary legislation, which usually specifies the principal restrictions that may be imposed and the kind of activities that may be controlled; such lists, however, can rarely be exhaustive, and should allow for additional measures to deal with unforeseen circumstances. The primary legislation should also provide for the enforcement of regulations by creating offences and prescribing penalties. Regulations need not be uniform throughout a marine protected area, and it is often more appropriate to designate internal zones with different regimes. These may be buffer zones where activities are less strictly controlled than in the principal area. Another approach, adopted by Australia in the Great Barrier Reef Marine Park Amendment Act 1988, is to classify the most sensitive sites as 'zoned areas' which must not be entered or used except for purposes that are expressly permitted; in unzoned areas of the park, the presumption is reversed, and only specified activities are prohibited. The degree of regulation that is legally permissible depends on the location of the marine protected area, and must conform to the principles of international law already described in section 3.1.1. Thus, in its land territory and internal waters, a State is free to impose whatever constraints it wishes; it may do likewise in its territorial sea, except that it must not infringe the right of innocent passage by foreign ships. In the exclusive economic zone, it may control pollution and interference with natural resources, but must otherwise permit freedom of navigation. On the continental shelf outside the exclusive economic zone, its powers are limited to protection of the sea-bed and sedentary species, and the control of pollution by dumping. These qualifications do not, however, apply to the coastal State's own subjects and vessels, which it may regulate anywhere in the world.

The matters that need to be controlled by regulations fall into three main categories. First, there are public rights to which people are ordinarily entitled; while there are inevitably variations between countries, the most common are the public rights of navigation and fishery in tidal waters. Second, there are public activities that are often tolerated, but which lack any legal basis; these may include such conduct as bait-digging and beachcombing. Third, there are private rights: for example, ownership of the foreshore or private fisheries. The removal or restriction of public rights in a marine protected area can properly be achieved by legislation, although the political will to do so may be more elusive, but it is important that the intention is unequivocally expressed, since there is a presumption of statutory interpretation that public rights should not be infringed without clear authority. Detrimental activities that are without legal foundation should in theory be easier to curtail, although they are often popularly assumed to have a lawful origin.

Private rights are more difficult to deal with. While the territorial sea-bed is usually State property, this is not always true of the foreshore between high and low water marks, and the adjacent coastal land is often privately owned; there may also be private fishing rights in inshore waters for such species as salmon and shellfish. Clear language is again required for the expropriation of private property rights, but political difficulties are more likely to arise and compensation is often necessary.

Another legal issue with serious political implications is the status of other government departments, statutory bodies and local authorities exercising administrative functions in the coastal zone. In common law countries, legislation is presumed not to bind the State unless there is an express statement to that effect; this means that the responsibilities of central government for other matters such as shipping, fisheries, waste disposal, mineral extraction and defence may take precedence over marine protected areas. Unless the primary legislation specifically resolves these conflicts, they are likely to undermine its practical effectiveness. The next section illustrates these problems by analysing the experience of the UK in the establishment of marine nature reserves.

3.2 IMPLEMENTATION – A CASE STUDY OF THE UNITED KINGDOM MARINE NATURE RESERVES PROGRAMME

The test of any piece of legislation lies in its implementation. The marine nature reserve (MNR) provisions of the Wildlife and Countryside Act 1981 came into force in November 1981 but the first reserve was not declared until November 1986 and the second in July 1990. Proposals for an MNR in the Menai Strait were submitted to the Welsh Office in March 1994. There are many reasons for the lack of progress in the marine reserve programme, not all of them reflecting on the quality of the legislation. The purpose of the following case study notes is to highlight the difficulties that have been encountered and hence to analyse the deficiencies in the legal provisions.

3.2.1 Sites selected for MNR status

Applications for MNR designations can only be made by one of the nature conservation agencies (English Nature, Countryside Council for Wales, Scottish Natural Heritage) which took over the role of the Nature Conservancy Council (NCC) following an administrative reorganization introduced in the Environmental Protection Act 1990. Their first task is to select sites suitable for designation. This depends on the necessary scientific information being available.

Although terrestrial sites have been reviewed for their nature conservation importance (Ratcliffe, 1977), the comparable Marine Nature Conservation Review did not commence until 1987. Nevertheless, the NCC believed that they had sufficient information to make an initial selection at the time when the Act came into force. Seven sites were on the list (Figure 3.1): Lundy and the Isles of Scilly in England; Skomer, Menai Strait, and Bardsey Island and the Lleyn peninsula in Wales; Loch Sween and St Abbs in Scotland. A further site for England, Lindisfarne, subsequently came under consideration. The marine nature reserve sections of the Wildlife and Countryside Act 1981 do not apply to Northern Ireland but similar provisions are contained in the Nature Conservation and Amenity Lands (Northern Ireland) Order 1985. Several sites are under consideration there and a firm commitment has been made by the Government to designate Strangford Lough. These sites are all relatively small, discrete areas of conservation value. Full details of the progress towards declaration of these sites as MNRs may be found in Warren (1989) and is summarized here in Figure 3.2.

Work on Lundy, Menai Strait, the Scillies and Skomer began in earnest as soon as the initial selection was made and for a time there was a race to see which one would be declared first. Serious progress on the Scillies halted when the Isles of Scilly Environment Trust was set up but survey work has continued. Lindisfarne was suggested as a possible English site when

Figure 3.1 Sites selected as possible statutory marine nature reserves.

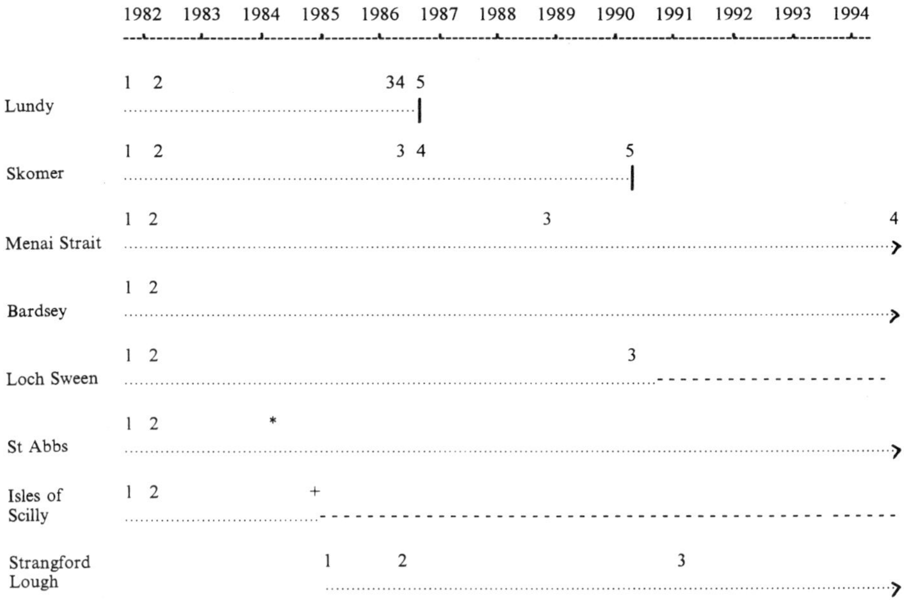

Figure 3.2 Progress towards declaration of marine nature reserves.
Key: --- Proposals unlikely to be pursued further.
1 Legislation in force.
2 Site selected.
3 Consultation paper issued.
4 Submission to Secretary of State.
5 Declaratory order made.
* Site declared to be a voluntary marine reserve.
+ Isles of Scilly Environment Trust set up.

extreme resistance to the Scillies reserve led to a decision to concentrate efforts elsewhere and wait and see how the Environment Trust developed. The Scillies are still referred to as one of the seven proposed MNRs, however, and Lindisfarne has never taken its place. Apart from biological survey work, carried out as part of the Marine Nature Conservation Review, there has been no drive towards the declaration of Lindisfarne.

In Wales most effort has been put into Skomer in south Wales and Menai Strait in north Wales. The third Welsh site, Bardsey Island and the Lleyn peninsula, is not yet ready to go to formal consultation.

In Scotland, the main thrust was directed towards Loch Sween. A Marine Conservation Officer was appointed to work on the designation but subsequently had to be diverted to deal with the far more urgent problem of fish farms in Scottish sea lochs. The formal consultation paper was not favourably received and no further progress has been made. The situation at St Abbs is being kept under review to see how the voluntary reserve operates before a final decision to proceed with statutory designation is made. Differences in the administrative arrangements mean that there is no need for a formal submission of proposed MNRs in Northern Ireland. Following the publication of the consultation paper for the Strangford Lough MNR there have been several public hearings but no formal designation has been made as yet.

3.2.2 Reasons for the delay

Administrative procedures

The legislation does not cover that part of the designation procedure that precedes submission to the Secretary of State. This is left to the country agencies acting under the direction of their lead ministry (the Department of the Environment in England, the Welsh Office and the Scottish Office). The Wildlife Division of the Department of the Environment issued procedures for pre-designation in March 1983. These are summarized as follows:

1. Site selection.
2. Identification of controls necessary to achieve objectives of reserve.
3. Consultation with government departments to obtain agreement in principle.
4. Discussions with Department of the Environment/Welsh Office/Scottish Office on any problems arising under (3).
5. Preparation of site map giving details of reserve boundaries and boundaries of any areas requiring separate byelaws.
6. An evaluation of the byelaws in three categories: (i) those already existing; (ii) those to be made by the NCC; (iii) those which may be made by other agencies, and consultation, in confidence, with the agencies involved.
7. Consideration of whether the International Maritime Organization should be asked to declare any part of the reserve to be an area to be avoided by shipping.
8. Preparation of consultation document.
9. Full consultation. Consultees are to be encouraged to make representations at this stage and not wait until a draft order has been made.
10. Evaluation of response and resolution of problems.
11. Submission to appropriate Secretary of State including full details of any outstanding objections where the assistance of or intervention by the Secretary of State is requested.

Approximate indications as to the expected duration of the consultation stages were included. Thus the preliminary discussions with government departments were expected to take one month and the consultation period following publication of the consultation document was optimistically set at three months. In hindsight these time periods seem ridiculously short and comparison with Figure 3.2 shows the true picture. The misjudgement appears to have been made through a complete failure to appreciate the fact that consultations have a habit of becoming open-ended unless there is some controlling mechanism to curtail them. The procedures were accompanied by interpretative guidelines to assist the NCC. These included an instruction to reject any site where there were clearly going to be important and largely irreconcilable differences of opinion. The NCC was expected to undertake thorough consultations with all interested parties and the expectation was expressed that any major objections would be resolved prior to submission to the Secretary of State.

In practice there are three stages in the designation process. Having made a preliminary assessment of a site as a potential MNR, the agency discusses its proposals with a wide range of interested parties such as people using the site, conservationists, local and central government departments etc. with the dual aims of publicizing the idea of a reserve and obtaining feedback on its suitability for such designation. At the same time further site surveys may be carried out, either by the NCC or under contract to them, in order to substantiate the scientific justification for MNR status.

The second stage starts with the publication of a consultation document by the agency. This explains why the agency wishes to make an application for MNR status for the site and goes on to show the geographical extent of the reserve and to explain the proposed byelaws. The precise format of the document has changed somewhat in the light of experience. The latest one to be published, that for Menai Strait, contains far more scientific information. Previous consultation documents had concentrated on the practical implications of the reserve and referred the reader to other sources for data to justify the proposal. The original Menai Strait consultation document was published at the same time as a much shorter leaflet answering the questions most commonly raised at public meetings. This had a much greater print run than the formal consultation document and was distributed widely to the general public. Readers of the consultation papers are invited to comment on the proposals to the agency. There then follows a period of negotiation in which the agency tries to resolve any objections made. In the case of the Menai Strait proposals a second consultation document was published taking into account comments made during the first formal consultation exercise.

The final stage involves a further period of consultation both before and after publication of a draft order. The agency submits an application to the Secretary of State. This must be accompanied by a copy of byelaws proposed for management of the reserve in accordance with section 36(2). The

Secretary of State may then make an order declaring the MNR. The procedure for making such orders is laid down in detail in Schedule 12 but the Act does not cover the process by which the Secretary of State makes a decision as to whether to proceed with a draft order or not.

Thus, although the agencies do not have the final say in deciding whether or not a site should become an MNR, the bulk of the preparatory work falls on them. In particular, they are burdened with the task of having to consult with a wide range of people and organizations and must further seek to resolve any objections. It is implicit in the Department of the Environment's administrative guidelines that applications are only to be made once all, or all major, objections have been overcome. Furthermore, procedural convenience would appear to take precedence over scientific desirability. There is no authority for these directions in the Act itself. According to the wording of section 36 it is merely incumbent on the agency to furnish the Secretary of State with copies of the proposed byelaws along with the application. The Act does not say that the agency must obtain a near consensus agreement for the proposals. The matter does not arise in connection with other habitat protection designations because these are in the direct power of the agencies. It is unclear why it is necessary to involve the Secretary of State at all if objections must be sorted out before submission. Schedule 12 of the Act contains provisions for holding inquiries into the desirability of a designation where objections have been made to the draft order. The inclusion of this mechanism suggests that Parliament had intended the Secretary of State to make a decision having reviewed the evidence for and against the designation rather than act as little more than a rubber stamp.

Central government action following submission of an application has, in fact, been rather different in respect of the submissions made so far. This may be partly because different ministries have been involved but it also relates to the timing of the applications. The informal consultation stages regarding Lundy were being undertaken during a period in which the House of Commons Select Committee on the Environment was considering the effectiveness of the Wildlife and Countryside Act and Parliament was considering the Wildlife and Countryside (Amendment) Bill which included a clause to amend the marine nature reserve sections of the Act. The unexpected delay in making a first declaration was, therefore, given high profile in Parliament and led to government announcements to the effect that further delays would not be tolerated (see, for example, Waldegrave, 1986). As a consequence of this, there was active ministerial involvement in the informal consultation stage to ensure that the legislation could be seen to be working. From the time that the formal consultation document was issued in June 1986 to the date of the order declaring the reserve was just five months. The Secretary of State for the Environment was clearly very keen to get the designation made as quickly as possible so a speedy passage through departmental channels was ensured. The boundaries of an inner area in which trawling was to be prohibited were changed in response to last-minute objections made by local fishermen who had not previously been consulted. There was no consultation with other parties on these changes and the draft order did not actually refer to them as such. In other words, a decision must have been made that the consultation stage should not be re-opened.

The Welsh Office involvement in the Skomer reserve proposals has been much greater. Protracted negotiations by the NCC failed to resolve objections raised in response to the consultation paper which was issued in the summer of 1986. The NCC, therefore, decided to proceed with the submission nonetheless in the expectation that the Secretary of State would resolve the issue either by exerting pressure before proceeding with publication of a draft order or by holding an inquiry following publication. In the event no further progress was made and the Welsh Office returned the submission to the NCC in July 1987. A further year of negotiations and compromises was necessary before the NCC was able to resubmit the application in November 1988. Despite its best endeavours it was unable to overcome all of the objections made. This time the Welsh Office did engage in discussions to resolve the outstanding problem concerning the wording of one of the byelaws.

Further problems arose, however, in the precise wording of the order. There was some concern that the Lundy order had been made rather hurriedly and it was the intention of the Welsh Office that everything should be properly correct this time. There followed several months in which the NCC byelaws were reworded to the satisfaction of both Welsh Office and NCC and the draft order was not published until March 1990. Of the representations made very few contained any objections to the declaration and those that did were satisfactorily resolved without the need for further hearings. The order declaring the Skomer and Marloes peninsula Marine Nature Reserve was accordingly made in July 1990.

The Lundy consultation paper was sent out to about 130 bodies. In the case of Skomer, 220 organizations and individuals were on the original mailing list but further copies were sent out. The Menai Strait package was issued to over 1500 consultees so it is not surprising that some of the responses were in opposition to the reserve. Nevertheless, of the 129 formal responses received by the Countryside Council for Wales (CCW) only 21 were opposed and only a few of these made specific objections that could not be resolved and remained outstanding when the submission was made. Although the Government has pledged its continued support for the MNR programme (see, for example, Anon 1994b), there is no indication, as yet, as to how the Welsh Office will respond to the submission.

This section has concentrated on the consultation procedure itself as a cause of delay. The problem is compounded by the nature of the disagreements involved as is described below.

Limits to control

Proposals for marine nature reserve legislation were not included in the original Wildlife and Countryside Bill and were only accepted by the Government with reluctance. The government approach to the subject has always been that reserves should only be created where there is voluntary approval for the proposals from those most affected by them (Warren and Tydeman, 1989). As a consequence various activities and institutions were excluded from the effects of the legislation during its passage through Parliament. The result is that only a very few activities can be regulated under MNR byelaws. Not surprisingly this discrimination has led to much ill-feeling and is one of the prime reasons for objections to date.

The Act provides that neither an MNR designation nor byelaws made in respect of it shall interfere with the exercise of any functions of a relevant authority, any functions conferred by or under an enactment (whenever passed) or any right of any person (whenever vested). Relevant authorities include local authorities, statutory water undertakers, navigation authorities and others who have responsibilities in the coastal zone. For the most part their activities are unlikely to come into conflict with the interests of a marine reserve. The list also includes sea fisheries committees, however, and these have direct responsibility for the conservation of fish. The role of sea fisheries committees in marine nature reserves has given rise to problems in relation to MNR proposals for several sites. Institutions not included in the relevant authority list may, nevertheless, be protected by this exclusion if their powers and responsibilities derive from statute as, for example, is the case with the Crown Estate Commissioners who have a duty to manage the Crown Estate, including most of the foreshore and the sea-bed, under the Crown Estate Act 1961. Private individual rights, such as private rights to take shellfish, are also protected.

The scope of byelaws made under the Act is further restricted in that they cannot be used to control the exercise of any right of passage by any vessel except a pleasure boat where limited control, but not total exclusion, is possible. There are also exclusions for actions taken in an emergency, for discharges from vessels and for anything done more than 30 m below the sea-bed, i.e. submarine mine shafts where entry to the mine is from land outside the reserve.

Sea fisheries committees are established under the Sea Fisheries Regulation Act 1966 and are responsible for making byelaws for the conservation of commercial species of fish. They come under the control of the Ministry of

Agriculture, Fisheries and Food rather than the Department of the Environment. The original government clauses for marine nature reserves expressly excluded fish from the types of animal that could be protected by the byelaws but this measure was withdrawn following strong opposition. The Act thus allows the agencies to make byelaws to protect animals of any description which, on the face of it, should include fish. The NCC was advised, however, that it does not have powers to make byelaws prohibiting the taking of commercial species because this would amount to interference with the functions of the relevant sea fisheries committee. Instead, the agencies must persuade the committee to make its own byelaws. This is far from satisfactory. Fishing is often the one activity that needs to be controlled but there may be some difficulty in convincing the sea fisheries committee that this is the case. Although these committees do not represent the interests of local fishermen as such, their conservation remit is interpreted from a fisheries viewpoint and not an environmental one. The passing of the Sea Fisheries (Wildlife and Conservation) Act 1992 has improved matters by imposing a duty on the committees to have regard to the conservation of marine flora and fauna in the exercise of their duties. Nevertheless, fishermen are notoriously suspicious of conservationists and the problem is not helped by the fact that there is a distinct antipathy between fishermen and divers.

Opposition from the Devon Sea Fisheries Committee held up the Lundy MNR for over a year. They were asked to make byelaws banning dredging and bottom trawling throughout the reserve to protect the soft sediment communities and to ban the setting and lifting of pots within 100 m of the Knoll Pins where there was a rich but fragile fauna that could easily be damaged. In practice there was no trawling within the boundaries of the proposed reserve and the only fishing was a small amount by line and potting, the former being largely to supply bait for the latter. Nevertheless, the committee refused to cooperate until the Ministry of Agriculture, Fisheries and Food took an active interest in the promotion of the reserve. As a result a compromise was reached. Devon Sea Fisheries Committee byelaws prohibit spear-fishing throughout the reserve and prohibit trawling and the use of nets in specified areas of the reserve only. Gill-netting is permitted in one specified area; elsewhere it is forbidden without written consent. There are also restrictions on potting in the vicinity of the Knoll Pins.

Even greater resistance was shown by the South Wales Sea Fisheries Committee in respect of the Skomer proposals. The NCC wished to ban scallop dredging because of the enormous damage this causes to benthic communities. More than a year after negotiations had started, the committee challenged the assumption that damage resulted from such activities. In order to proceed further, the NCC was persuaded to conduct an experimental survey on the effects of dredging at this site (Bullimore, 1985), after which the committee was prepared to draw up the necessary byelaw prohibiting the use of any dredge or beam trawl within the reserve boundaries. The committee also readily agreed to make byelaws prohibiting the taking of sea fish, including shellfish, by divers and prohibiting spear-fishing.

These byelaws were opposed by the diving organizations who felt that they were being victimized. Spear-fishing had never been common at Skomer and had virtually ceased in 1970 so they saw no need for it to be singled out for a total ban while angling was not to be prohibited. Similarly they objected to the ban on taking shellfish by divers as there was no ban on potting for shellfish. The divers' objections were amongst those outstanding when the NCC made their initial submission to the Welsh Office.

When the application was turned down the NCC was able to resolve most of the objections by adopting a compromise position (for further details see Gibson, 1988 and Warren, 1989) but was not prepared to make any concessions on the fisheries byelaws. Negotiations throughout most of 1988 concentrated on attempts to overcome this obstacle and eventually the sea fisheries committee was persuaded to replace the byelaws prohibiting dredging and the taking of fish by divers with a new one which prohibited the taking of fish by any method except by the use of pots, traps, or rod and line. It was hoped that this would allay the divers' fears of discrimination but this was not the case and the divers wanted the byelaw amended so that taking by hand would also be allowed. The NCC considered that this would remove one of the main purposes of the byelaw and made its resubmission with the dispute unresolved. There then followed a period of several months in which the Welsh Office negotiated for a form of words favourable to both divers and fishermen and the byelaws were redrafted as follows:

- No person shall use in fishing for sea fish any fishing dredge or any beam trawl within the [specified] area.
- No person shall fish for, take or land any scallop of the species *Pecten maximus* or of the species *Chlamys opercularis* from the [specified] area.

This compromise has the effect of protecting the most vulnerable species in the benthic community while at the same time allowing divers and fishermen alike to take other species such as lobsters. Ideally, the NCC would have preferred a total ban and has reserved the right to come back on this one if necessary in the future.

So, not only was the designation delayed by about two years because of this dispute, but the final result falls short of providing the protection that is desirable for the reserve. Even when agreement had been reached, there was further delay because of the mechanism for confirmation of the byelaws. The NCC or agency byelaws can be confirmed by the Department of the Environment (or, in this case, the Welsh Office) but byelaws made by sea fisheries committees must be approved by the European Commission before being confirmed by the Ministry of Agriculture, Fisheries and Food or the Welsh Office. Although it is unlikely that approval would be withheld, it does mean a further period of bureaucratic delay.

Problems over fisheries byelaws are likely to arise in connection with any proposal for a marine reserve in England or Wales and there are already signs of a major confrontation between divers and fishermen in respect of the Bardsey Island and Lleyn peninsula proposal. In Strangford Lough, damage due to dredging is one of the major threats to the site.

3.2.3 Statutory interpretation

Section 37(1) of the Act enables the agencies to make byelaws 'for the protection of any area designated as a marine nature reserve'. There then follows a list of specific purposes in section 37(2) which is stated to be without prejudice to the general purpose. When preparing the case for the Skomer MNR, the NCC considered that it was necessary to prevent anchoring in a small part of the reserve area in order to protect a sea grass bed. One of the specific purposes concerns the control of 'entry into, or movement within, the reserve' (section 37(2)(a)(i)) but the NCC was advised that anchoring did not come within this and that it would have to prohibit access completely in order to prevent anchoring (Gibson, 1988). Opposition from yachtsmen led the NCC to abandon their plans for such a byelaw.

There was also opposition, this time from conservationists, to proposals to make byelaws for the protection of seabirds nesting on the cliffs at Skomer because these were not considered to be inside the reserve. The dispute illustrates a fundamental problem with habitat protection legislation in Great Britain, i.e. the lack of powers to control activities outside the reserve which have an adverse effect on conservation interests within the reserve. Skomer Island is a national nature reserve so the seabirds are protected from interference from the land. The National Parks and Access to the Countryside Act 1949 empowered the NCC to make byelaws controlling shooting in the vicinity of a reserve but there are no legal controls over other activities. The only way of protecting cliff nesting seabirds from disturbance from the sea, for example by fast boats moving too near, would have been to use the MNR legislation on the assumption that the cliffs were within the reserve boundaries. In the event, the NCC withdrew its plans to proceed with the byelaw and chose to use a voluntary code of practice instead.

Menai Strait contains isolated populations of the king rag (*Neanthes* = *Nereis virens*), a worm which is highly prized as bait. The CCW has decided not to seek an all-out ban on collection of this species but has instead proposed that collection should be prohibited in specified sanctuary zones. The council is not clear, however, as to the best way of achieving this end. King rags are clearly animals but are they commercial species that would come under control of the sea fisheries committee? At one stage it was intended to make a CCW byelaw controlling bait digging in the sanctuary zones and a complementary North Western and North Wales Sea Fisheries Committee byelaw to regulate shellfish collection in the same area but the CCW remain in some doubt as to whether a byelaw prohibiting the killing, taking, etc. of any animal in the reserve would include animals taken for use as bait (NCC, 1988; CCW, 1992). A recent court case finding that digging for lugworms (*Arenicola marina*) as bait was ancillary to the public right to fish does nothing to clarify the matter. This example illustrates further the problems of dividing responsibility for making conservation byelaws between two regulatory bodies.

Section 37(2)(b) of the Act enables the agencies to issue permits for carrying out activities that would otherwise be unlawful under the byelaws and allows them to attach conditions to these as specified in the byelaws. The Lundy byelaws thus enabled the NCC to grant permits subject to such conditions as it saw fit. The Welsh Office considered that such a provision might be *ultra vires* and the Skomer byelaw refers to specified conditions instead. It was felt that otherwise the NCC and, subsequently, the CCW might be in a position to attach conditions that undermined byelaws made by other authorities which could, therefore, be *ultra vires* because they might be interpreted as an interference with the functions of a relevant authority. The need to draft specific yet comprehensive conditions led to the declaration of the Skomer reserve being delayed for several months while a satisfactory set of conditions was drawn up.

3.2.4 Deficiencies in the legislation

Marine nature reserves are supposed to be at the top of a pyramid of devices for habitat protection. As such it should be possible to control all activities within a reserve. The legislation clearly fails to achieve this aim. The difficulties encountered in trying to control fisheries have already been referred to but there are also limitations in respect of other user groups.

For example, attempts to give powers to the NCC to control navigation were treated with extreme caution by the Government who feared problems over established rights of passage. (For further comments on the Government's attitude on navigation see Gibson, 1980). In the end it was conceded by the Government that it would be in order to control pleasure boats. This was considered satisfactory for the purposes of protecting a marine reserve as these are the only types of vessel likely to be there. A last-minute onslaught by the Royal Yachting Association, incensed at being singled out for control, led to the powers being weakened so that pleasure boats could only be prohibited from particular parts of the reserve at particular times of the year.

There is a further problem concerning the meaning of 'pleasure boats' which are not defined in the Act. Although the Government was quite clear that the term included private boats and boats hired for pleasure purposes, it is not certain that this is legally correct (Gibson, 1984) although there have been no problems to date. Byelaws controlling navigation were not considered necessary for the Lundy MNR. In the case of Skomer there is a byelaw restricting the speed of any boat within a large part of the reserve for part of the year which presumably applies only to pleasure boats.

The failure of nature reserve legislation to control activities taking place outside a reserve has been mentioned above with reference to national nature reserves. The consequences are potentially more disastrous for marine nature reserves where there are no natural boundaries to contain organisms or even the water itself. Around Lundy, for example, there has been a considerable increase in potting since the reserve came into being (Willcox, 1988). Although this occurs outside the reserve, the protected area is so small that it is likely that this will be detrimental to the reserve. There is also the general problem of pollution from some outside source affecting a reserve. It was decided not to include express measures to control pollution on the grounds that it was inappropriate to treat reserves in isolation. A better solution might have been to legislate for buffer zones around reserves where certain activities could be controlled as and when necessary.

3.2.5 Conclusions

The UK's marine nature reserve provisions would thus seem to be rather weak and ineffectual with no powers to control some of the potentially most hazardous activities. A number of amendments have been suggested with a view to strengthening the legislation, for example, by giving ministers powers to overrule the protection given to relevant authorities as was proposed in amendments to the Wildlife and Countryside (Amendment) Bill. The official view of the NCC, however, was that the legislation is the best that could be obtained. It was feared that anything stronger would have led to such hostility amongst certain user groups that valuable sites might be destroyed before designation was possible. There is also the problem of policing a reserve to enforce any stronger byelaws.

Obviously it is better to obtain the cooperation of all parties involved. Nevertheless, the legislation and the manner in which it is implemented carry the need for agreement too far. Given the nature of UK marine reserves as small sites in which there are selective controls, it is inevitable that there will be tensions and local resentments. One way of easing the problem might be to go for much larger reserves consisting of a very wide buffer zone, which might be managed using voluntary codes of conduct affecting all users, surrounding discrete reserve areas with strict controls as appropriate (see, for example, proposals for a new system of marine protected areas in Warren and Gubbay, 1991). Education is probably the key to the whole problem. If the public at large are in favour of marine nature reserves then it matters less that a particular user group is opposed. The NCC took a step in this direction with its proposals for the Menai Strait reserve. Unlike the two existing reserves, Menai Strait is close to a large human population whose numbers increase considerably during the summer months. Many of these people visit the reserve area. If the CCW has been able to convince them of the desirability of the reserve, this will go a long way towards its successful declaration.

Prospects for the designation of any further MNRs are uncertain. The Government's third annual report on its environmental strategy (Anon, 1994b) pledges to proceed with work on the proposed Lleyn peninsula MNR. The Biodiversity Action Plan (Anon, 1994a) is non-committal. It acknowledges that mechanisms for designation have sometimes proved cumbersome and notes that there are two MNRs in existence with two more (presumably Menai Strait and Bardsey and the Lleyn peninsula) under consideration but makes no comments on the likelihood, or even desirability, of further progress. It does, however, make a commitment to create mechanisms for the effective protection and management of key wildlife areas in the marine environment.

Government thinking is undoubtedly coloured by the need to comply with the EC directive on the conservation of natural and semi-natural habitats and of wild fauna and flora (92/43/EEC). This requires the establishment of a network of sites, to be called Special Areas of Conservation (SACs), representative of key habitats and species. Several marine habitats and species are included. Designated sites must be protected from deterioration and disturbance and management plans are to be drawn up where this is necessary to achieve the protection. If any significant adverse changes are sanctioned by government (on the grounds of overriding public need) there must be some sort of compensatory measures.

The Government's consultation paper on implementing the directive (Department of the Environment and the Welsh Office, 1993) considered that the MNR provisions would not, as they stand, form an appropriate basis for implementation. It also rejected an earlier government proposal, made in a draft circular issued in February 1992, to set up a series of Marine Consultation Areas (MCAs). These were to have been formal but non-statutory designations designed to ensure that conservation was taken into account when new developments were being considered.

The Conservation (Natural Habitats, &c) Regulations 1994, which implement the directive, contain elements of both MNRs and MCAs. Relevant authorities (including the nature conservation agencies, sea fisheries committees and other statutory bodies) are empowered to make a management scheme for a European marine site, under the direction of central government if necessary. Furthermore the nature conservation agencies may make byelaws under section 37 of the Wildlife and Countryside Act 1981 subject to the same restraints as apply when such byelaws are made for MNRs.

At first sight the Regulations appear muddled and, because of the defects of the byelaw-making powers, inadequate to provide the required degree of protection. Some of the problems of cooperation may be avoided by the issue of directions for management schemes but, with so many relevant authorities and with more than one central government department involved, it is by no means certain that there will be agreement as to specific management requirements.

What is certain is that the Government will be directing its attention to marine SACs in the immediate future. Whether this means the demise of the MNR programme is difficult to predict. The list of proposed SACs is not publicly available but it is likely that the existing MNRs will be included. This does not mean that MNRs can simply be subsumed within the SAC network, however. Some sites may be worthy of MNR status on a UK basis but fail to meet the European status. More importantly, MNRs may be designated for educational and research purposes and, therefore, have a wider remit than SACs. Part of the justification for the Menai Strait proposed MNR, for example, is the opportunities a reserve would provide for raising public awareness of marine conservation. Whether the Government is prepared to recognize the continued value of MNRs is, however, debatable.

REFERENCES

Anon (1994a) *Biodiversity. The UK Action Plan*, Cm. 2428, HMSO, London.

Anon (1994b) *This Common Inheritance. The Third Year Report*, Cm. 2549, HMSO, London.

Bullimore, B. (1985) *An Investigation into the Effects of Scallop Dredging within the Skomer Marine Reserve*. Report to the NCC by the Skomer Marine Reserve Subtidal Monitoring Project, 39 pp.

Burke, W.T. (1981) National legislation on Ocean Authority Zones and the Contemporary Law of the Sea. *Ocean Development and International Law*, 9, 289–322.

Countryside Council for Wales (1992) *Menai Strait Proposed Nature Reserve*, Consultation Document.

Department of the Environment and the Welsh Office (1993) *Implementation in Great Britain of the Council Directive on the Conservation of Natural Habitats and of Wild Fauna and Flora (92/43/EEC), 'The Habitats Directive'*. Consultation Paper.

Freestone, D. (1990) Specially Protected Areas and Wildlife in the Caribbean – The 1990 Kingston Protocol to the Cartagena Convention. *International Journal of Estuarine and Coastal Law*, 5(4), 362–382.

Gibson, J. (1980) Marine conservation and the right of navigation. *Lloyd's Maritime and Commercial Law Quarterly*, 2, 203–207.

Gibson, J. (1984) Marine nature reserves. *Journal of Planning and Environmental Law*, 699–706.

Gibson, J. (1988) Marine nature reserves in the United Kingdom. *International Journal of Estuarine and Coastal Law*, 3(4), 328–339.

IMO (1991) *Ships' Routeing*, 6th edition, International Maritime Organisation, 1991.

Law of the Sea Bulletin 23 (1993) United Nations Division for Ocean Affairs and the Law of the Sea, June 1993.

Lyster, S. (1985) *International Wildlife Law*, Grotius, Cambridge, 470 pp.

Nature Conservancy Council (1988) *Menai Strait Proposed Marine Nature Reserve*. Consultation document, NCC Wales, 47 pp.

Ratcliffe, D.A. (1977) *A Nature Conservation Review*, Cambridge University Press, Cambridge.

United Nations (1983) *Convention on the Law of the Sea, Montego Bay, 10 December 1982*. UN Sales Publication No E.83.V.5, New York.

Waldegrave, W. (1986), speaking in response to a Parliamentary question, *96 H C Debs*, c 311–312 (19 February 1986).

Warren, L.M. (1989) *Statutory Marine Nature Reserves in Great Britain. A Progress Report*, World Wide Fund for Nature, Godalming, 41 pp.

Warren, L.M. and Gubbay, S. (1991) *Marine Protected Areas, A Discussion Document*, World Wide Fund for Nature, Godalming, 12 pp.

Warren, L.M. and Tydeman. C.F. (1989) A voluntary approach to statutory MNRs in the UK, in *Coastal Zone '89. Proceedings of the Sixth Symposium on Coastal and Ocean Management, Charleston July 11–14, 1989*, (eds O.T. Magoon, H. Converse, D. Miner *et al.*), American Society of Civil Engineers, New York, 2650–2664.

Willcox, N. (1988) Lundy Marine Nature Reserve – one year on. *Marine Conservation*, Spring 1988.

4

Involving the community

Susan Wells and Alan T. White

> It is becoming obvious that the key to protecting a cherished landscape lies within the communities that call it home. (IUCN, 1993).

4.1 INTRODUCTION

There has been increasing interest in recent years in community-based management of natural resources or, where the government is responsible for management, in public participation. Community involvement can range from autonomous management by the community to some form of shared responsibility with state agencies. This trend was in part stimulated by the World Conservation Strategy (UNEP/IUCN/WWF, 1980) which, in promoting the link between conservation and development, represented a major change from the rigid preservationist attitude of many early conservationists (Hanks, 1984). Early protected areas were usually set up by government agencies with the intention of keeping people out. On land this is at least theoretically possible with fences and barriers; it is also possible to move people out of a planned protected area. However, as population pressure and demand for land and resources has increased, this philosophy has become inappropriate and increasingly difficult and expensive to implement (Hough, 1988; McNeeley and Miller, 1984). Hence in many countries there has been growing emphasis on involving local people at some, if not all, levels of protected area establishment and management (IUCN, 1993; Kemf, 1993; Wells *et al.*, 1992).

The need for marine protected areas (MPAs) has been recognized comparatively recently, and the potential conflict between local people and MPA establishment became apparent very quickly. The characteristics of the sea mean that marine resources cannot be fenced off; furthermore, in many parts of the world they are considered to be available to everyone. Many MPAs have therefore been set up as multiple-use areas, zoning for different activities is commonplace (Chapter 6) and in some areas the concept of

Marine Protected Areas: principles and techniques for management. Edited by Susan Gubbay.
Published in 1995 by Chapman & Hall, London. ISBN 0 412 59450 1.

community-based management has received early acceptance (e.g. Castaneda and Miclat, 1981; White and Savina, 1987; White *et al.*, 1994).

Community involvement in MPA management can reinstate the idea of marine resources as 'common property'. This term is used for resources that are collectively owned by a community, in contrast to 'open access' resources which are open to exploitation by any individual or group. Coastal communities in many countries had, and in some cases still have, customary laws or unwritten regulations to prevent individuals maximizing their private gains at the expense of community interests (Berkes, 1985; Bromley and Cernea, 1989). In such cases, for example in Oceania and parts of North America (e.g. Ontario and Quebec), these traditions and customs can lead to sustainable use of their resources, whether intentionally or indirectly (Johannes, 1978; Berkes, 1985).

Before the arrival of the industrial society, communities living near marine resources used them and often depended on them for food and other materials. Deterioration of these resources could therefore be quickly recognized. This direct dependence on the sea has now disappeared in many areas and with it the community's sensitivity to the consequences of resource damage and depletion. In subsistence cultures, resource deterioration may still be felt but the community often lacks the means to deal with it, in the face of outside commercial pressures and increasing population (Kenchington, 1988). Furthermore, population growth, technological change, commercialization and loss of community control over resources have led to widespread destruction of traditional marine tenure systems. In Oceania, pressure from colonial governments to abandon traditional conservation laws so that trade surpluses would be generated (Johannes, 1978) and the establishment of a western system of public ownership of subtidal lands (Holthus, 1989) often triggered the breakdown of these systems. The appearance of open access systems quickly follows, leading to overexploitation and the often cited 'tragedy of the commons', or perhaps more appropriately 'tragedy of open access' (Berkes, 1985; Bromley and Cernea, 1989).

Traditional management practices often involved regulating use of particular sites, and efforts are now being made to incorporate these practices into modern MPA management. Even where such practices have been lost for many years, or are not known to have existed, community involvement can result in the development of a similar sense of responsibility and concern for a marine and coastal area. This chapter examines a number of aspects of community involvement with particular reference to experience from the Philippines. There are clear links with the role of enforcement, education and interpretation, the involvement of indigenous peoples and voluntary conservation in MPAs, each of which is discussed in some detail in Chapters 7–10.

4.2 WHY COMMUNITY INVOLVEMENT?

Where people are dependent on their adjacent marine resources for their livelihoods, the establishment of an MPA is likely to have a significant impact on their lives and, inevitably, results in a reaction from the community. The challenge to managers of MPAs is to channel this response into support for the project. Involving the community can reduce the conflict which often arises in MPA establishment where the best sites for conservation are also those most economically valuable (Kenchington, 1988). The need for community involvement has been neatly put by Kaza (1988): 'With involvement comes understanding, with understanding comes public support and commitment.' There are numerous cases, particularly in developing countries, which show that the establishment of an MPA by a government agency, without consultation with local people or without their subsequent involvement in its maintenance, is ineffective in protecting marine resources (White, 1986a; UNEP/IUCN, 1988/89).

Community involvement can have important benefits for both the local people and the protected area. Sharing of economic benefits from the MPA can increase income for local people, for example from user fees (e.g. from tour operators and entrance fees) and visitor facilities (accommodation, transport, food, guides, etc.). Improved employment opportunities may arise, both in and outside the MPA, through the growth of services such as hotels and restaurants. A successful MPA may result in improved yields in local fisheries (White, 1988; Alcala, 1988). Although not dependent on community involvement *per se*, such benefits can have an enormous impact where the community is involved. Important ingredients of community involvement are incentives and sanctions that will influence the behaviour of individuals who live in or depend on an area (Bromley and Cernea, 1989; McNeely, 1988).

For MPA managers, community involvement can facilitate enforcement of regulations, as local people will understand and accept their purpose more readily. The community can assist or even be responsible for enforcement, thus reducing costs to government agencies. Where there are financial constraints local people, especially in developed countries, can be mobilized to help develop interpretive programmes and assist with education (Kaza, 1988).

4.3 WHO IS THE COMMUNITY?

The type of community using or adjacent to an MPA has an important bearing on the way community involvement develops. At one extreme, local people may be predominantly subsistence-level fishermen. At the opposite extreme, there may be a diverse array of interests: commercial and recreational, as well as subsistence, fishermen; yachtsmen, marina owners and people servicing the sailing industry; tourists and tourist operators, both national and international; those involved in agriculture, industry, transport and other coastal activities; and, in some cases, residents who live on the coast primarily for its scenic value. An MPA will impinge on the interests of these people in widely different ways.

Representatives of the tourist and recreation industries within a community may be supportive of MPA establishment. Many MPAs attract large numbers of divers, snorkellers, boaters and recreational fishermen, and the economic benefits of their activities is well documented (Thorsell and Wells, 1990; van't Hof, 1985). Hotels, dive operators and other tourist operators are increasingly approaching local or national governments with requests to protect sites that are attractive to their clients. However, promotion of an MPA for tourism may be unpopular if other local residents object to an increase in visitor numbers. For example, in the 1970s there was initially opposition to the proposed Looe Key National Marine Sanctuary, Florida. It was feared that more visitors would lead to even greater reef damage; less altruistically, commercial operators feared that diving and recreational fishing might be curtailed (Clark *et al.*, 1989).

Local fishermen are often opposed to MPAs, assuming that there will be an adverse economic impact on their livelihoods. Conservation efforts in the Houtman Abrolhos archipelago in Western Australia were initially vociferously opposed by the local rock lobster fishermen, who maintained that their activities were not affecting marine resources and that the creation of a reserve would cause depletion of stocks elsewhere (Wright *et al.*, 1988). On the other hand, if the local community has a tradition of marine resource management, people may welcome the resuscitation of such ideas and be predisposed to community-based controls. Opposing views may develop within a community. In the UK, for example, differences in opinion between divers and fishermen have contributed to the difficulties in establishing statutory marine reserves (Warren and Tydeman, 1989).

4.4 LEVELS OF COMMUNITY INVOLVEMENT

Clearly no single set of guidelines will cover all situations for involving a community in MPA management. The nature and extent of involvement will depend on the aims, purpose and characteristics of the MPA, the history of its establishment, the resources it contains and their value, and the structure of the local community. In a few instances community involvement may be unnecessary, as in MPAs established in extremely remote areas such as some of the protected Pacific atolls. Four ways in which communities can become involved are considered in more detail:

1. Community-based reserves, where there is maximum involvement.
2. Involvement through the existence of traditional regulatory customs or their reinstatement.
3. Community involvement in government establishment of MPAs through a collaborative arrangement.
4. Biosphere reserves.

4.4.1 Community-based MPA management

Community-based MPAs are those set up and managed by people living in the immediate vicinity. Such MPAs may have no formal government endorsement initially, but are often subsequently supported by government (national or municipal) legislation, and receive input from government programmes. As well as protecting biological resources and preventing deterioration of the marine environment, they often have a more specific purpose related to the community. This may be improvement of fishery yields, as in the case of the Philippine community-based reserves (White, 1988) or increasing public awareness of the marine environment so as to foster better general environmental management of the sea and coast, as in the case of UK voluntary marine nature reserves (Warren and Tydeman, 1989; see also Chapter 10). Community-based MPAs function best where the MPA is small and fairly well defined, the local community has a close dependence on or interest in marine resources, and the benefits of the MPA are clear.

Community-based MPAs have been particularly successful in the Philippines. They were first developed by Castaneda and Miclat (1981) in the coastal municipalities of Sagay and Guindulman in Negros, where advice and educational programmes were made available to villagers, who then drew up local regulations and enforcement procedures for activities on reefs. Subsequently, the Marine Conservation and Development Program of Silliman University assisted in the establishment of community-based reserves around the small islands of Balicasag, Pamilacan and Apo (Savina and White, 1986; White, 1989b; White and Savina, 1987). Each of these was set up by the community, following an educational programme by project workers who lived on the islands and introduced the concept of the reserves. Each MPA is managed by a marine management committee, selected from the local community, and is formalized and given legal backing at the municipal level through a municipal order. The boundary generally encircles the island, at about 500 m from shore. Damaging fishing methods, such as dynamite, poisons, spear-fishing and *muro-ami* (hitting the reef with rocks and sticks to scare fish into a net – this causes considerable damage to any coral and is extremely efficient at catching most of the fish in the area), are prohibited within the reserve. Buoys and signs demarcate part of the reserve as a fish sanctuary, within which all fishing and collecting of marine organisms is prohibited (section 4.6). This system is now being used elsewhere in the Philippines, such as the Lingayen Gulf (McManus *et al.*, 1988), the Central Visayas (Alix, 1989) and San Salvador Island in Zambales (Christie *et al.*, 1994).

Resorts and tourist operators in many parts of the world are starting to set up informal MPAs to protect popular dive and snorkelling areas. These often have a primarily commercial objective but if well managed can benefit the marine environment in general. In Fiji, for example, a resort manager and the local village, from whom the two resort islands were leased and which had traditional tenure over the area, collaborated in setting up an MPA. Fishing controls were implemented through the traditional village system. The resort contributed in more modern ways, such as installing a sewage system that prevented pollution and prohibiting damaging tourist activities. It benefited from healthy, diverse reefs which attracted tourists, and the local people benefited from the employment generated (UNEP/IUCN, 1988/89).

The disadvantage of community-based management is that if non-local people choose to abuse a MPA, the local people have no authority to prevent it. There are no means of enforcing regulations other than peer pressure, education and persuasion. Changes in personnel can also affect policy. The first Philippine municipal reserve, Sumilon Island, had serious problems when a change in local government resulted in the new mayor authorizing commercial fishing within the sanctuary. This led to major, but temporary, destruction of the coral reef (Alcala, 1988; White 1986b).

4.4.2 Community involvement through traditional customs

Traditional ownership of marine resources, through reef tenure or fishery user rights, still exists in several areas of the Pacific and can be used to improve MPA protection (Holthus, 1989). Throughout much of the region, clans or large communal groups own and control the rights to fish on their adjacent reefs. In some countries traditional tenure may have evolved to ensure a sustainable yield of reef resources, to minimize conflicts and to distribute resources effectively (Ruddle and Johannes, 1989). Elsewhere, for example in Papua New Guinea and Indonesia, there may have been a political or social basis (Polunin, 1984; Chapman, 1987). These systems have outwardly disappeared or been weakened in many countries, for example in Hawaii, the Marianas and American Samoa, but may still offer a strong base for community involvement, as the concept of enforcement of regulations by local people may still be understood. Sometimes the traditional customs may be sufficient in their existing form; in other cases it may be desirable to recognize them openly and provide a modern legal foundation to strengthen them.

The Wildlife Management Area (WMA) system in Papua New Guinea, where over 90% of the land is still under traditional ownership, illustrates how traditional rights can be legalized. A WMA is usually established at the request of the traditional landowners, who approach the Department of Environment and Conservation following identification of some particular wildlife management problem. The area is designated under the Fauna (Protection and Control) Act of 1978. The boundaries are established after discussion between the local people and appropriate government departments; the management regulations are drawn up by the community and are approved by the Department of Environment and Conservation. Generally, traditional forms of hunting by the community are permitted. The regulations are enforced by the community with assistance from government officers as required. Enthusiasm for the system is so great that there are too few government staff to provide the necessary advice, assistance and follow-up. Most WMAs are terrestrial but several have been established for turtles and dugong. For example, Maza WMA, covering an area of 1886 km^2, was set up in 1979 for the management of dugong. Representatives of the six villages involved make up the committee and oversee the enforcement of regulations which prohibit the taking of dugong except by local people. There are restrictions on the size and numbers taken and, for each individual taken, the committee is paid a royalty which is put back into the management of the WMA (Asigau, 1989). Chapter 8 describes the issues surrounding the involvement of indigenous people in MPAs in some detail with reference to the Great Barrier Reef Marine Park.

4.4.3 Community involvement in statutory protected areas

Government-controlled protected areas are still often thought of in terms of the traditional concept of national parks. As originally defined by IUCN, national parks required the elimination of exploitation in and occupation of the area. The only concession to human use and economic activities tended to be tourism, the benefits from which generally went to the national economy rather than to local people. Not surprisingly, the establishment of national parks often had consequences as damaging as those which might have occurred had the area been left unprotected. In 1982, the Bali Declaration, drawn up at the World Parks Congress, called for an increase in 'local support for protected areas through such measures as education, revenue sharing, participation in decisions, complementary development schemes adjacent to the protected areas and, where compatible with the protected area's objectives, access to resources' (McNeely and Miller, 1984). This has resulted in a less restrictive concept of national parks and reappraisal of the relationship between government-controlled protected areas and local people in many countries (Hough, 1988; IUCN, 1993). Nevertheless, several government protected area programmes have attracted criticism for their failure to involve local communities satisfactorily. This can be a result of underfunding and staff shortages, preventing the necessary community development work, lack of appropriate training for park staff, or conflict between those who still hold to the rigid 'national park' concept and those who aspire to multiple-use management. However, there are encouraging signs that this situation is changing. Many MPAs have education of the public as one of their main objectives, and often an MPA becomes an important focus of the community, even if there was little initial community participation.

Government involvement is often essential in the protection of large areas, where the disparate interests of different communities and users may conflict, and at sites where the economic benefit of the MPA is not immediately evident to local people (e.g. protected areas for endangered species). Equally, community involvement in government-controlled management is also now seen to be essential (Bromley and Cernea, 1989). The Bunaken National Marine Park, near Manado, Sulawesi, established in 1991, is a good example of government collaboration with local park residents. The park covers 89 000 ha of land and marine waters, has five islands and includes major areas of coral reef, sea-grass and mangrove ecosystems. Island residents have been traditionally dependent on fishing and are now starting to benefit from tourism. The management plan is being developed through collaboration between local people and local government officials. The size of the area and the complexity of its management, combined with threats from outside sources such as non-local fishermen using damaging methods, tourism and pressure from coastal development in Manado, mean that some form of collaboration with government is essential if the planned community-based management is to be feasible (White, 1992).

A zoning plan, developed with the villagers, formalizes traditional fishing areas for reef fish and invertebrates and includes strict protection zones where no exploitation is allowed. Scuba-diving and tourist visitation areas are also zoned. The island communities organize committees for management and protection of the sanctuary areas with the understanding that they will benefit from improved reef-fish yields. The management scheme includes the participation of several non-governmental organizations from Manado working with the National Park Office on matters of community-level education.

The need for co-management of this kind is also well illustrated by a project underway in the Soufriere region on the west coast of St Lucia. Although the marine environment was designated a Marine Reserve Area in 1986, and plans were made in 1987 to develop a National Park to cover both the terrestrial and marine components, no active management was implemented. As conflict between user groups and stakeholders increased (a result of the growing importance of the area as a scuba-diving destination and increasing demand for fishery resources) it became clear that effective management would be feasible only if all those with interests in the area were involved. A process of negotiation and participatory planning was therefore initiated, which culminated in the formation of the Soufriere Regional Development Foundation which will oversee management of the area and represent all the stakeholders, and the publication of a preliminary agreement on the use and management of marine and coastal resources for the area, to be termed the Soufriere Marine Management Area. At the same time, a coral reef monitoring programme has been set up, which also involves many of those who make their living from the marine environment and in particular the dive operators. This cooperative approach to planning and decision-making should facilitate the production and implementation of the final management plan for the area (Smith, 1994; Soufriere Regional Development Foundation, 1994).

Fagatele Bay National Marine Sanctuary in American Samoa illustrates how traditional management can be incorporated into government programmes. The site was nominated in 1982, and the planning stages involved wide consultation and local participation from the village concerned. Traditional subsistence fishing and gathering of fish and invertebrates from the reef flat at low tide are permitted and projects are being developed to involve communities in management (Thomas, 1988).

Community involvement is increasingly found at the consultative stage and in educational and interpretive programmes. The Great Barrier Reef Marine Park in Australia provides an example. The Marine Park Authority promotes community involvement and generates public support which leads to long-term national pride in the Park. Local people are consulted during the development of zoning plans and involved in a variety of public education, interpretive and recreational activities. Although this is not full community participation, public support generated in this way is essential as in the long run it leads to greater government support for the MPA (Kelleher, 1986).

In the United States, increasing emphasis is being put on citizen participation, in the belief that active and informed citizens will be one of the greatest allies in the formation and establishment of MPAs. National Marine Sanctuaries are only designated following consultation with citizens' groups and the general public as well as Congress and other federal and state agencies. Marine conservation non-governmental organizations, such as the Center for Marine Conservation, work closely with local scientists and conservationists to mobilize support for proposed sites. The sanctuary programme calls for multiple use and allows recreation, fishing, education and other compatible uses. Local citizen support groups are formed to help with fund-raising, education and interpretive programmes (Cava and Power, 1989; Kaza, 1988; Clark *et al.*, 1989).

In British Columbia, Canada, the yachting community has become involved in the management of the 40 or so marine parks in the province. The Ministry of Parks has set up a marine park 'host' system whereby local yacht clubs or individual yachtsmen are encouraged to anchor in a marine park for a set period to provide other visitors with information about the marine park scheme (Turner, 1990).

4.4.4 Community involvement in biosphere reserves

Biosphere reserves are established under the UNESCO Man and the Biosphere (MAB) Programme and are designed to provide the framework necessary for the involvement of local people. The concept of a biosphere reserve is very different from that of the traditional form of national park, in that human activities are considered to have a constructive role in environmental protection. Biosphere reserves aim to combine the functions of preservation of biological diversity, research, education and training within a single site, through a zonation system. This comprises a core area of minimal human disturbance, a buffer zone where activities such as research, traditional land or marine use, recreation, tourism and education are allowed, and a transition area outside the reserve in which the work of the biosphere reserve is applied directly to the needs of the local communities. The aim is that biosphere reserves should be 'demonstration sites of harmonious, long-lasting relationships between man and the natural environment' (Robertson Vernhes, 1989) and in recent years it has been recognized that this concept is particularly relevant to the marine and coastal environment (Price and Humphrey, 1993).

At present, few biosphere reserves have full community participation. For example, islanders still lack involvement in the Virgin Islands Biosphere Reserve, although some have been given training in research and resource management and others have helped on reef projects. In 1988, a group of local residents formed the 'Friends of the Virgin Island National Park' specifically to encourage better communication and interaction between the park and the local community, particularly in the transition area (Rogers, 1989).

One of the best coastal/marine examples of a biosphere reserve is the Sian Ka'an Biosphere Reserve in Mexico, which has a Council of Representatives to represent the needs of the local people living in the reserve (Carillo-Barrios-Gomez and Herrman-Martinez, 1989). Most interestingly, community-based management of marine resources existed in this area before the Reserve was established. Under Mexican Federal law, lobster, conch and shrimp may only be fished by cooperatives which are allotted particular areas. Many such ventures have failed, but the two lobster cooperatives in the Sian Ka'an area have been successful. Each fisherman has his own territory where he establishes and maintains lobster shelters, and also controls access to other marine resources. Territories can be bought and sold among cooperative members, and are sufficiently well recognized that they can be inherited. The cooperatives have agreed on certain limitations, such as closed seasons and permissible equipment, and each fisherman is responsible for protecting his own territory. Penalties for poaching include banning from the cooperative and confiscation of equipment by the cooperative. The combination of incentives such as rights and peer support, and disincentives such as confiscation and peer pressure, seems to be successful (Miller, 1986).

4.5 HOW TO INVOLVE THE COMMUNITY

Community involvement can take place at several stages in the establishment and management of an MPA. Specific management practices may be unique to the communities employing them but the techniques of encouraging local support are widely applicable. Hough (1988) and Kaza (1988) list a number of ways that statutory MPAs can involve the community; White (1989b) and Gubbay (1986) outline the procedures for establishing community-based MPAs in the Philippines and Britain respectively. White *et al.* (1994) describe six case studies of community involvement in coral reef management, including examples relevant to MPAs. Some or all of the following general principles apply to most situations.

4.5.1 Recognition of a need

The recognition of a need for an MPA often comes from outside the community, for example from a local conservation organization or scientists aware of the biological significance of an area. In Britain, the rapid growth of recreational scuba-diving in the 1970s gave rise to concern among conservationists that fragile nearshore marine communities could be damaged and led to the establishment of Voluntary Marine Nature Reserves (VMNRs). In the Philippines, scientists identified potential sites for community-based reserves and then initiated an education programme to introduce the concept of marine reserves to the local people (White and Savina, 1987). Once a few successful reserves have been established, communities will often approach local conservation groups, scientists or government agencies for assistance in establishing a reserve. In the case of statutory MPAs, local people can play an important role in stimulating government action. If the initiative comes from the government, consultation with the local community at the earliest possible stage will enhance the likelihood of success of the project.

4.5.2 Discussion with interested parties/integration with the community

This stage is often delicate but, if successfully carried out, it may greatly reduce the time taken over consultation and negotiations with different user groups and community interests. In Philippine community-based reserves, trained community workers move into the area for a period of up to six months. They introduce the idea of the MPA, meet community leaders, attend local meetings and generally become accustomed to the local culture and traditions (White and Savina, 1987). In Britain, this stage involves formal meetings with all those involved in a proposed VMNR as well as more general public meetings when all issues are aired and a consensus of public opinion is obtained. For statutory MPAs, it is often more effective if initial consultation takes place through third parties such as local environmental organizations and politicians, and if local or traditional meeting places and halls are used for meetings rather than government offices (Hough, 1988).

Traditionally, park staff are trained in biological rather than social skills, but it is now recognized that training in negotiation and diplomacy is also required (Hough, 1988; Wright *et al.*, 1988). The community workers appointed to Philippine community-based reserve projects are fully trained in this way. Their success led McManus *et al.* (1988) to propose the term 'Environmental Community Organizers' (ECOs) for such people. They need to have (or to develop quickly) an intimate knowledge of local conditions, as well as a sound knowledge of community psychology, economics and sociology. Ideally such community workers should also be capable of basic marine resource monitoring and fishery assessment. The appointment of such a person can be of enormous benefit, particularly for the establishment of community-based MPAs (White *et al.*, 1994).

4.5.3 Baseline studies and monitoring

Data collection and scientific studies for the development of the management plan and its later evaluation are normally started at the beginning of an MPA project. In most cases this activity is carried out and organized by the project staff, but there is often scope for community involvement. In particular, local people may be able to provide important data, such as seasonal variations in fish abundance, distribution of different marine ecosystems, resource use and meteorological information (Smith, 1994). In Philippine community-based MPAs, feedback to the local people on the results of scientific work has helped to generate interest in and acceptance of the MPA.

4.5.4 Education

This is a process that continues at all stages of establishment and management of an MPA. Initially, the emphasis will be on explaining to local people the basic principles of marine ecology and resource management, its relevance to the immediate situation and the relevance of the proposed MPA to the community. This can be achieved through small group meetings and an occasional more formal presentation with slides, posters and other visual material. Non-formal methods are often best, permitting participation and interaction and encouraging personal contact (Cabanan and White, 1981; Kaza, 1988). Where possible, local people such as teachers or students should be used in the educational programme; they will often be able to put the 'message' across in the most appropriate way for the community. As individuals become enthusiastic about the project and involved in it, their assistance can be solicited to spread the word and help with the education process. It can be particularly helpful if local community leaders visit similar projects.

4.5.5 Core group building and formalization of the reserve

In most successful community involvement projects, a key step is the formation of a committee. In the Philippine community-based marine reserves, the project community worker helps to set up a marine management committee, which initiates drafting of the municipal ordinance. In British VMNRs, a steering group is often set up to direct the initial stages of reserve establishment. This group selects a reserve committee which includes representatives of all the users' interests and is responsible for management of activities within the reserve. In Papua New Guinea, a committee is established as a focus for the establishment of a Wildlife Management Area (WMA). In Sri Lanka, a local government and community coordinating committee has been formed to assist with management of the Hikkaduwa Marine Sanctuary which was established in 1979 but has not been successfully managed until recently (White and Samarakoon, 1994). The establishment of these committees is extremely important as it gives real responsibility and power to the community for management of the protected areas. Public meetings may be sufficient in some cases but generally do not lead to the same sense of community participation (Hough, 1988). A committee can become involved in a wide range of activities such as drafting and approval of the management plan, MPA regulations, any subsequent changes to them and any other aspects of the management of the area.

As the local community may represent a variety of different interests, smaller groups of individuals may also be expected to form (Hough, 1988) to deal with particular issues. In the Philippine reserves, subgroups have been set up to organize the construction of park facilities such as visitor centres or to carry out education and liaison activities.

4.5.6 Enforcement

The core group or committee (variously named management committee, marine management committee, etc.) will play an important role in enforcement. There is good evidence that enforcement by the community, through peer group pressure and local disincentives, is more effective and less costly than government enforcement. However, as mentioned earlier, the latter may be necessary where an area is large, where there are conflicting uses and interests and where violations are from non-locals. The subject of enforcement is discussed in more detail in Chapter 7 with reference to arrangements for Looe Key and Key Largo National Marine Sanctuaries in the United States. In statutory MPAs, there should be a long-term commitment to hire and train local people, not only as seasonal labourers and wardens and to run facilities, but also to become senior park staff (Hough, 1988). In many cases, a combination of local enforcement and government assistance may provide the best solution. In Philippine community-based reserves and Papua New Guinean WMAs, the management committee is responsible for liaising with outside enforcement agencies as needed.

4.5.7 Strengthening community involvement

This is most effectively achieved through positive feedback and publicity of the success of the MPA. Favourable comments from visitors and interaction between visitors and the community, through guided tours, hotels and restaurants run by local people, help to confirm the value of the MPA in the eyes of the community. Exchange visits with other communities involved with MPAs, perhaps combined with training sessions, can foster interest and pride.

Monitoring of key biological parameters should be carried out from the beginning, and the results passed back to the community. The increase in fish population and fishery yields in Philippine community-based reserves had a major impact and encouraged further local participation (Alcala, 1988; White, 1988; White, 1989b). Similar results in Looe Key National Marine Sanctuary greatly increased public support and dispelled some of the earlier opposition (Clark *et al.*, 1989).

Involvement of schools, local non-governmental organizations and the local media, provision of printed materials, newsletters, badges and T-shirts, and special events can all help to foster support. Local people can help to build visitor facilities and construct exhibits, signs and trails and this encourages them to take an interest in their maintenance. A popular strategy is to establish adoption programmes ('Adopt-a-trail', -reef, -animal, etc.)

which enable people to support some aspect of the MPA either financially or in kind. For example, the Adopt-a-trail scheme in the Gulf of Farallon National Marine Sanctuary in the United States enables local people to undertake to maintain certain trails (Kaza, 1988). Other ideas are described in Chapter 9 on the role of education and interpretation in MPAs.

For community-based MPAs, the maintenance of contact with relevant government agencies and non-governmental organizations is essential. These organizations must be willing to provide support when problems (such as those involving politics, legal issues and planning decisions) arise that the community lacks the skills or authority to handle. This assistance must be provided through the channels set up by the community and in a way which is acceptable to it. A community may lose its self-esteem if a government agency heavy-handedly takes over some aspect of the MPA which is normally dealt with by local people.

4.6 CASE STUDY – BALICASAG ISLAND MUNICIPAL MARINE PARK

The Philippines is plagued with marine resource management problems because of its densely populated coast and the dependence of its people on fishery and coral reef resources. It has thus proved a good country in which to experiment with community-based resource management. Local people are a part of most, if not all, marine conservation programmes in the country as there are few coastal areas which are not inhabited or accessible. Balicasag Island Municipal Marine Park provides a case history of complete community involvement in the establishment and management of an MPA, and is thus a good example of the so-called 'bottom-up' approach to management.

4.6.1 Description of site and history of project

Balicasag is a small island of about 30 ha in the southern part of the Philippines. The adjacent marine environments were surveyed in 1976 by the Bureau of Fisheries and Aquatic Resources and in 1978 a proposal was drawn up for a national marine park. This was never acted upon, but it drew the attention of conservationists, divers and other tourists, particularly shell collectors, to the island's rich marine environments. Balicasag is surrounded by a fringing coral reef with high coral cover in some parts and a steep drop-off along two-thirds of the reef. The reef is now well known as a good scuba-diving area, and the island has been famous for some time for the valuable semi-precious shells which are collected from depths of 100 m or more in nearby waters (White and Dobias, 1990). The island community consists of some 600 people who, prior to the establishment of the reserve, made their living from fishing and shell collecting.

In the early 1980s, the Marine Conservation and Development Program (MCDP) of Silliman University was developing a community-based marine reserve on the nearby island of Pamilacan. The project staff made several trips to Balicasag to build interest in the principles of marine conservation. These initial contacts revealed great enthusiasm and, shortly after, the Balicasag islanders asked the MCDP for assistance in setting up their own MPA.

The project was started in 1985 and proceeded rapidly, due to the interest shown by the local people. The community worker assigned to the island was received favourably and helped to set up a marine management committee. The community agreed on a zoning scheme comprising two areas:

- a sanctuary area covering 8 ha and extending 550 m along the south-west shore in which fishing and collecting of marine organisms is strictly prohibited but snorkelling, diving and swimming are permitted;
- a marine reserve surrounding the entire island to a distance of 500 m from shore (thus covering the entire coral reef) within which traditional and ecologically sound methods of fishing are permitted and any form of destructive fishing is prohibited.

These zones were established by a municipal ordinance written by the marine management committee. Management guidelines were drawn up by the community and have generally been enforced. The community continues to be involved in all aspects of park management, including the installation of signs and buoys, patrolling the area and informing tourists about the park and use of anchor buoys. The only complicating factor has been when the Philippine Tourism Authority (PTA) set up a 'backyard tourism project' which established a small resort on the island for the benefit of the community. As mentioned below, this small resort and its management has tended to take credit that is due to the community for management of the marine reserve.

4.6.2 Stages in involving the community

Many of the principles discussed in section 4.5 applied in the case of the Balicasag MPA. Five stages in the project, several of which overlapped and built on each other, could be identified.

Integration of the project community worker into the community and baseline data collection

Because of the ready acceptance of the community worker by the islanders, this stage proceeded rapidly and took only about two months. The community worker collected baseline data on the marine resources, the island and the community itself, and oriented the people to the aims of the project. She was careful to listen to the needs and perspectives of the islanders. The marine survey was an important part of the integration process as the community wanted information on the condition of their coral reef. The attraction of the island to scuba-diving tourists had shown the community that there was an economic aspect to the aesthetic value of the reef, which was thus worth conserving.

Education

Educational activities took place throughout the project but were less time-consuming than in some projects because of the enthusiasm of the community for marine conservation. Nevertheless they wanted to learn more about how an MPA would function and how they would benefit from it. In particular, they hoped that the MPA would lead to an increase in visitor numbers and thus increase the sales of the shells they collect and process. The shells are taken mainly from beyond the 500 m limit of the reserve. Shell collecting is not entirely consistent with the principles of an MPA, particularly as little is known about the biology and life history of the species involved. Sustainable harvest levels are also completely unknown. However, perhaps the establishment of an MPA in the area will lead to proper monitoring of this fishery and ultimately to research to provide the data necessary for developing a sustainable fishery plan. It is hoped that the education that the islanders received in the course of the establishment of the MPA has sensitized them to the problems of over-exploitation, particularly of a resource which is so little understood and apparently ‘rare’.

Formal and informal educational methods were used. There were also classes in mat weaving, shell craft and tourism which were aimed at developing ecologically sound alternative livelihoods.

Organization of a resource management core group

A marine management committee was formed following the success of activities such as the building of a community education centre, installation of signs and buoys, teaching, etc. which showed the community that it was capable of carrying out projects. The committee was made up of people who had worked on the community education centre and representatives from the local government unit on the island, the ‘barangay council’. It became the focus for the MPA project and provided the framework for management activities. It was also able to resolve problems. For example, the committee initially met resistance from several families on the island who had been involved in illegal fishing in the past and were not enthusiastic about marine conservation, but in due course this was resolved. There was further resistance when the ordinance was presented to the municipal government council, as one of the council members was involved in illegal aquarium fish collection using poison. He tried to veto the ordinance but was overruled by the mayor, who supported the MPA proposal.

Building responsibility around beneficial projects

As the benefits of the MPA started to become apparent to the islanders, so their willingness to take responsibility for management was strengthened. Of particular importance was the noticeable increase in fish abundance in the sanctuary area, which led to an increase in snorkelling and diving visitors and also improved the fish catch within the reserve area. A second survey, one year after establishment of the MPA, showed that coral reef fish diversity and abundance had both significantly increased. This naturally pleased the local people and reinforced their management efforts (White, 1988). Their sense of responsibility was also enhanced when the marine management committee became confident enough to warn off illegal fishermen from other islands. This is no mean achievement in the Philippines, where fishermen may carry guns and operate illegally without fear of apprehension by government authorities.

Formalizing the community role in MPA management

Two events contributed to this. Firstly, Silliman University initiated a research project on giant clams (Tridacnidae) in the sanctuary zone. Secondly, the Philippine Tourism Authority (PTA) chose Balicasag as a pilot site for its 'backyard tourism' programme, which has the aim of encouraging small tourism enterprises at the community level by people who would not have the resources or know-how to start a small tourist business. A small resort was built on the island in 1987 and is operated by the PTA in collaboration with the islanders, some of whom are employed in the resort. It is intended that some of the profits from the resort will go to marine conservation efforts on the island. Unfortunately, the resort has been designed in a somewhat sophisticated way, making it difficult for the islanders to manage. This means that they have been unable to participate fully in its operation. This intrusion also highlighted the fragility of the community management regime because of some conflicts with the new resort management which have forced the island community to coexist with the resort in a manner not of their choosing. Nevertheless, the resort has emphasized the importance of a healthy coral reef for tourism and has shown the community how tourism can be a viable alternative to fishing as a livelihood (White and Dobias, 1990).

Recognition of the MPA and the efforts of the community by organizations such as Silliman University provided an enormous boost to the community. Indication that the MPA was fully accepted also came from the fact that local mechanisms which are legal or accepted at the community government level were used to help in resolving difficult management problems such as political disputes, problems of law enforcement and local organizational changes.

4.6.3 Continuity of management

Prospects for the future management of the MPA on Balicasag Island are good as long as tourism continues to provide a reasonable income. The interest shown in the MPA by Silliman University and the town government provides a further incentive and encouragement for continuing management. The PTA involvement, although potentially supportive, has been a mixed experience from the perspective of the local community. The visible improvement in the health of the reef is important: in 1989, the reef had denser fish populations and better coral cover than at any time in the past 12 years (White and Dobias, 1990). As the community worker was careful to ensure that the community did not become dependent on her for guidance in the project, the islanders have had confidence in their actions even after her departure. This independence also bodes well for the future management of the MPA.

4.7 CONCLUSIONS

Community involvement has become a widely accepted part of protected area management in recent years, as conservation has evolved to take account of human needs as well as wildlife protection. There are now several cases which illustrate how MPA management can be improved through the participation of local people. Involvement leads to an understanding of the principles involved, making enforcement easier and increasing support from the local community, whether financial, political or practical. The community may benefit financially through increased employment, sharing in the income generated by the MPA, or through improved fishery yields resulting from resource management. The overall effect is to reinstate the idea of marine resources as 'common property', for which the community is responsible.

It is difficult to generalize about how communities should be involved. Methods and types of involvement will vary according to the structure of the community, the policies and structure of national and local government agencies and the characteristics of the MPA. Community-based MPAs are appropriate in some situations, particularly where the site is small and well defined and where the community is closely involved with and dependent on the marine environment (e.g. through fishing or tourism). In most cases these MPAs will ultimately require some form of government support and collaboration to provide the necessary authority to deal with infractions by outsiders or conflicts between different user groups.

In other situations it may be more effective for the government to establish an MPA, particularly if the area is large and involves a number of different communities, if there are already conflicts between users, or if there are few people with an interest in the marine resources or their management. However, government reserves will generally be far more effective if local people are involved wherever possible in the processes of establishment and management. Such collaborative approaches are enshrined in the concept of biosphere reserves and successful examples of these can provide useful models. Community involvement may be particularly appropriate and easy to implement in regions which have, or have recently had, some form of traditional customary law regulating the use of marine resources, as local people are already familiar with the principles involved. The same general principles seem to apply in most situations, including community involvement in terrestrial protected areas (IUCN, 1993).

Firstly, the concept of a new protected area should be introduced very carefully to the community and where possible in such a way that it may seem that the community itself had the initial idea. Information gathered about the area should be passed back to members of the community so that they develop a greater understanding of its significance. Secondly, the benefits of protected areas must be made clear to people and their needs should be addressed. The process of education should be seen as on-going and not limited just to the initial stages. Thirdly, in many cases, particularly where the aim is to establish a community-based reserve, the appointment of a community worker, trained in both resource management and community skills, may be enormously beneficial and sometimes essential. A further important principle is the formation of a committee to oversee the development and management of the protected area. This is essential for community-based reserves but may also be of value in government protected areas, where the views and needs of local people need to be coordinated and presented to the relevant authorities.

Where possible, the community should be responsible for enforcement, allowing traditional disincentives and peer pressure to operate. Finally, a protected area needs to be managed in such a way that its future survival is guaranteed beyond the early stages when community support may mainly reflect the novelty value of the project. Long-term support from the community will be encouraged through the benefits, financial or otherwise, that accrue from a protected area, from community pride and from the recognition bestowed on it by other communities and organizations. The protected area must itself be economically viable, either on a self-sustaining basis or through adequate government support. Community involvement should not be seen by governments as a way of cutting costs. Involving the community can cost as much in time and money as running a protected area directly from a government agency, but the guarantee of success is much greater and in certain situations may be the only means of ensuring long-term stable management.

REFERENCES

Alcala, A.C. (1988) Effects of protective management of marine resources on fish abundance and fish yields in the Philippines. *Ambio*, **17**, 194–199.

Alix, J.C. (1989) Community-based resources management: the experience of the Central Visayas Regional Project 1, in *Coastal area management in south-east Asia: policies, management strategies and case studies*, (eds T.T. Chua and D. Pauly), ICLARM Conf. Proc. **19**, 185–190.

Asigau, W. (1989) The Wildlife Management Area system in Papua, New Guinea. *Proc. 4th South Pacific Conference on Nature Conservation and Protected Areas, Vanuatu*, **2**, 134–146.

Berkes, F. (1985) Fishermen and 'the tragedy of the commons'. *Env. Cons.* **12**(3), 199–206.

Bromley, D.W. and Cernea, M.M. (1989) *The Management of Common Property Resources: Some Conceptual and Operational Fallacies*. World Bank Discussion Paper, No. 57, The World Bank, Washington DC, 66 pp.

Cababan, A.S. and White, A.T. (1981) Marine conservation program using non-formal education at Apo Island, Negros Oriental, Philippines. *Proc. 4th Int. Coral Reef Symp.*, **1**, 317–321.

Carillo-Barrios-Gomez, E. and Herrman-Martinez, H. (1989) A new biosphere reserve model in Mexico, in *Proc. Symp. on Biosphere Reserves, 4th World Wilderness Congress, 1987, Colorado*, (eds W.P. Gregg, S.L. Krugman and J.D. Wood), 223–233.

Castaneda, P.G. and Miclat, R.I. (1981) The municipal coral reef park in the Philippines. *Proc. 4th Int. Coral Reef Symp.*, **1**, 283–285.

Cava, F. and Power, D.M. (1989) Environmental information, the public and the Channel Islands National Marine Sanctuary. *Coastal Zone '89*, **3**, 2665–2677.

Chapman, M.D. (1987) Traditional political structure and conservation in Oceania. *Ambio* **16**(4), 201–205.

Christie, P., White, A.T. and Buhat, D. (1994) Issues and applications: community-based coral reef management on San Salvador Island, the Philippines. *Society and Natural Resources*, **7**, 103–117.

Clark, J.R., Causey, B. and Bohnsack, J.A. (1989) Benefits from coral reef protection: Looe Key reef, Florida. *Coastal Zone '89*, **4**, 3076–3086.

Gubbay, S. (1986) *Conservation of Marine Sites: a Voluntary Approach*, Marine Conservation Society, UK.

Hanks, J. (ed.) (1984) *Traditional life-styles, conservation and rural development*, Commission on Ecology Papers No. 7, IUCN, Gland, Switzerland.

Holthus, P.F. (1989) Marine protected areas and conservation of marine resources:

the South Pacific situation – status and strategies. *Proc. 4th South Pacific Conference on Nature Conservation and Protected Areas, Vanuatu*, **3**, 177–181.

Hough, J.L. (1988) Obstacles to effective management of conflicts between national parks and surrounding human communities in developing countries. *Env. Cons.*, **15**(2), 129–136.

IUCN (1993) *Parks for Life; Report of the IVth World Congress on National Parks and Protected Areas*, IUCN, Gland, Switzerland, 260 pp.

Johannes, R. (1978) Traditional marine conservation methods in Oceania and their demise. *Ann. Rev. Ecol. Syst.* **9**, 349–364.

Kaza, S. (1988) Community involvement in marine protected areas. *Oceanus*, **31**(1), 75–81.

Kelleher, G. (1986) Managing the Great Barrier Reef. *Oceanus*, **29**, 13–19.

Kemf, E. (1993) *The Law of the Leotho; Protection Indigenous Peoples in Protected Areas*, Sierra Club Books, San Franciso, 296 pp.

Kenchington, R.A. (1988) Managing reefs and inter-reefal environments and resources for sustained exploitive, extractive and recreational uses. *Proc. 6th Int. Coral Reef Symp., Australia*, **1**, 81–87.

McManus, J.W., Ferrer, E.M. and Campos, W.L. (1988) A village-level approach to coastal adaptive management and resource assessment (CAMRA). *Proc. 6th Int. Coral Reef Symp., Australia*, **2**, 381–386.

McNeeley, J.A. (1988) *Economics and Biological Diversity: Developing and Using Economic Incentives to Conserve Biological Resources*, IUCN, Gland, Switzerland, 232 pp.

McNeeley, J.A. and Miller, K.R. (eds) (1984) National Parks, Conservation and Development: the Role of Protected Areas in Sustaining Society. *Proc. World Congress on National Parks, Bali, Indonesia, 1982*, Smithsonian Institution Press, Washington DC.

Miller, D.L. (1986) *Technology, territoriality and ecology: the evolution of Mexico's Caribbean spiny lobster fishery*. Paper presented at Workshop on Ecological Management of Common Property Resources, 4th International Congress of Ecology, Syracuse, New York.

Polunin, N. (1984) Do traditional marine 'reserves' conserve? A view of Indonesian and Papua New Guinea evidence. *Senri Ethnology Studies* **17**, 267–283.

Price, A. and Humphrey, S. (eds) (1993) *Application of the Biosphere Reserve Concept to Coastal Marine Areas*. Papers presented at the UNESCO/IUCN San Francisco Workshop of 14–20 August, 1989, IUCN, Gland, Switzerland, 114 pp.

Robertson Vernhes, J. (1989) Biosphere reserves: the beginnings, the present, and the future challenges, in *Proc. Symp. on Biosphere Reserves, 4th World Wilderness Congress, Colorado, 1987*, (eds W.P. Gregg, S.L. Krugman and J.D. Wood), US Dept Interior, National Park Service, Atlanta, Georgia, pp. 7–19.

Rogers, C.S. (1989). Virgin Islands Biosphere Reserve: progress report, in *Proc. Symp. on Biosphere Reserves, 4th World Wilderness Congress, 1987, Colorado*, (eds W.P. Gregg, S.L. Krugman and J.D. Wood), 92–97.

Ruddle, K. and Johannes, R.E. (1989) *Traditional Marine Resource Management in the Pacific Basin: an Anthology. Contending with Global Change*, Study No. 2, UNESCO/ROSTEA, Jakarta, Indonesia.

Ruddle, K., Hviding, E. and Johannes, R.E. (1992) Marine resources management in the context of customary tenure. *Marine Resources Economics*, 7, 249–273.

Savina, G.C. and White, A.T. (1986) A tale of two islands: some lessons for marine resource management. *Env. Cons.*, **13**(2), 107–113.

Smith, A.H. (1994) Community involvement in coral reef monitoring for management in the Insular Caribbean, in *Collaborative and Community-Based Management of Coral Reefs, Lessons from Experience*, (eds A. White *et al.*), Kumarian Press, USA, 130 pp.

Soufriere Regional Development Foundation (1994) *Soufriere Marine Management Area: agreement on the use and management of marine and coastal resources in the Soufriere Region, St Lucia*, The Soufriere Foundation, Soufriere, St Lucia, 25 pp.

Thomas, W.J. (1988) Fagatele Bay: a sanctuary in Samoa. *Oceanus*, **31**(1), 18–24.

Thorsell, J. and Wells, S.M. (1990) A global overview of tourism activities in coastal and marine parks, in *Proc. 1990 Congress on Coastal and Marine Tourism, Honolulu*, (eds M.L. Miller and J. Auyong), National Coastal Resources Research and Development Institute, Newport, USA, **1**, 221–224.

Turner, M. (1990) *Community involvement in British Columbia's Provincial Marine Park Program*. Presentation at Congress on Coastal and Marine Tourism, Honolulu, May 1990.

UNEP/IUCN/WWF (1980) *World Conservation Strategy*, IUCN, Gland, Switzerland.

UNEP/IUCN (1988/89) *Coral Reefs of the World*, Vols 1–3, UNEP, Nairobi/IUCN, Gland, Switzerland, and Cambridge, UK.

van't Hof, T. (1985) The economic benefits of marine parks and protected areas in the Caribbean region. *Proc. 5th Int. Coral Reef Cong., Tahiti*, **6**, 551–556.

Warren, L.M. and Tydeman, C. (1989) A voluntary approach to statutory MNRs in the UK. *Coastal Zone '89*, **3**, 2650–2664.

Wells, M., Brandon, K. and Hannah, L. (1992) *People and Parks. Linking Protected Area Management with Local Communities*, World Bank/WWF/USAID, Washington DC.

White, A.T. (1986a) Marine reserves: how effective as management strategies for Philippine, Indonesian and Malaysian coral reef environments? *Ocean Management*, **10**, 137–159.

White, A.T. (1986b) Philippine Marine Park pilot site: benefits and management conflicts. *Env. Cons.*, **13**(4), 355–359.

White, A.T. (1988) The effect of community-managed marine reserves in the Philippines on their associated coral reef fish populations. *Asian Fish Sci.*, **1**(2), 27–41.

White, A.T. (1989a) The Marine Conservation and Development Program of Silliman University as an example for Lingayen Gulf, in *Towards Sustainable Development of the Coastal Resources of Lingayen Gulf, Philippines*, (eds G. Silvestre, E. Miclat and T.E. Chua), ICLARM Conference Proceedings **17**, pp. 119–123.

White, A.T. (1989b) Two Community-based Marine Reserves: Lessons for Coastal Management, in *Coastal area management in Southeast Asia: policies, management strategies and case studies*, (eds T-E. Chua and D. Pauly), ICLARM Conference Proceedings, Manila, **19**, pp. 85–96.

White, A.T. and Dobias, R.J. (1990) Community marine tourism in the Philippines and Thailand: a boon or bane to conservation? in, *Proc. 1990 Congress on Coastal and Marine Tourism, Honolulu,* (eds M.L. Miller and J. Auyong), National Coastal Resources Research and Development Institute, Newport, USA, 2, 453–459.

White, A.T., Hale, L.Z., Renard, Y. and Cortesi, L. (eds) (1994) *Collaborative and Community-Based Management of Coral Reefs. Lessons from Experience*, Kumarian Press, USA, 130 pp.

White, A.T. and Samarakoon, J.I. (1994) Special Area Management for Coastal Resources: A First for Sri Lanka. *Coastal Management in Tropical Asia*, **2**, pp. 20–24, Colombo, Sri Lanka.

White, A.T. and Savina, G.C. (1987) Community-based marine reserves: a Philippine first. *Proc. Coastal Zone '87, 26–30 May, Seattle*, pp. 2022–2036.

Wright, G., Hatcher, A.I. and Hatcher, B.G. (1988) Clarifying the impact of fishing activity on the reefs of the Houtman Abrolhos: results of a tandem approach between anthropology and marine science. *Proc. 6th Int. Coral Reef Symp., Australia*, **2**, 433–437.

5

Making a management plan

Richard Kenchington and Graeme Kelleher

5.1 INTRODUCTION

The best form of management plan for a marine or estuarine protected area depends on the reasons for which the protected area has been created. These should generally be specified in legal instruments which provide the basis and authorization for the establishment of the protected area being planned.

At the 17th General Assembly of the IUCN in 1988, a policy concerning marine and estuarine protected areas was adopted. This included a series of generic objectives which can cover the whole spectrum of responsibility and planning for the management of the marine resources of nations. The IUCN policy is based on the premise that marine environments and resources are so strongly linked over such great distances that conservation, and particularly the establishment of marine protected areas (MPAs), must be considered in the broad national context, or internationally.

The way in which these objectives may be achieved will vary from country to country and even between parts of countries. At one extreme is formal integration by a specified agency which has the responsibility of providing for conservation and sustainable multiple use of the marine environment as a multiple use management area incorporating zones of greater restriction such as Scientific Reserves or Strict Nature Reserves (IUCN type I) or National Parks (IUCN type II) for purposes of recreation, environmental appreciation and education. At the other extreme a national park or equivalent service has responsibility for some areas of the marine environment which are set aside and managed defensively as marine parks or reserves, while the remainder of the marine environment is managed within limits of informal policy coordination between competing agencies. The relationship between objectives and management strategies is discussed at more length by Salm and Clark (1984) and Kenchington (1990).

A range of situations may be covered by the term 'marine protected area' and there are three primary functions which should be served by management planning:

Marine Protected Areas: principles and techniques for management. Edited by Susan Gubbay.
Published in 1995 by Chapman & Hall, London. ISBN 0 412 59450 1.

- Area selection and definition.
- Management prescription – to specify the terms and conditions under which the area may be used or entered, consistent with the objectives of the area.
- Logistic planning – to define tasks for management officers, rangers, boatmen and others involved in management; and to specify personnel, training, equipment, finance and other resources required to achieve management.

In any particular system the term 'management plan' may cover one or more of these functions.

Site selection is discussed in Chapter 1 while this chapter concentrates on the processes and documents needed to establish management prescriptions and to involve and educate resource users and the broader community in planning and management. These processes do not occur in isolation so there will inevitably be overlap with other chapters.

The strategy of this chapter is to discuss the control mechanisms which can be used in management, to describe the phases of a programme to develop a management plan, and then to list and discuss briefly issues which need to be addressed in a tactical plan to implement management strategy. There is no particular case study but there is a complete annotated table of contents for a document or series of documents and attachments for covering the entire process of development of a hypothetical management plan. That table, presented by Kelleher and Kenchington (1992), results from cumulative development by a number of working groups. It started with a marine protected area planning workshop at the Third World congress on National Parks, held in Bali in 1982, reported by Salm and Clark (1984). It was extended by Kenchington (1984), and was further extended in an unpublished technical workshop on selection and management of marine and estuarine protected areas held in Australia in 1985 and by a working group of the Australian Committee for the International Union for the Conservation of Nature and Natural Resources (ACIUCN, 1986). It is presented not as a model which should be followed in precise detail, but as an attempt at a comprehensive listing of issues which should be considered from the point of decision that protection or management is desirable through to the point of implementation of management. Consideration will involve deciding whether each section or issue is significant or relevant in the context of a particular protected area proposal.

5.2 DEFINING THE OBJECTIVES OF MANAGEMENT

Establishing objectives for the management of the protected area is the most critical process in management planning. If those objectives are clear, appropriate and feasible, the prescriptions of the plan and resulting management should follow logically. To a large extent, the process of planning is one of achieving agreement on feasible objectives by the managers, resource specialists, those managed, and others legitimately concerned about management.

If the objectives are technically impractical (for example because the area is inadequate to guarantee survival of a viable breeding population of a key species, or because sufficient resources are not available) or socially unacceptable (for example because they fail to meet the subsistence needs of traditional inhabitants) the process is futile. The resulting plan, by creating the illusion of protection and management where none exists, is likely to do more harm than good. The IUCN policy, adopted at the 17th General Assembly in 1988, established a number of objectives which are recommended to all governments for incorporation in laws and policies for establishing a global, representative system of marine protected areas (Kelleher and Kenchington, 1992). The policy states that such a system should have the following objectives:

- To protect and manage substantial examples of marine and estuarine systems to ensure their long-term viability and to maintain genetic diversity.
- To protect depleted, threatened, rare or endangered species and populations and in particular to preserve habitats considered critical for the survival of such species.
- To protect and manage areas of significance to the life cycles of economically important species.
- To prevent outside activities from detrimentally affecting the marine protected areas.
- To provide for the continued welfare of people affected by the creation of marine protected areas.
- To preserve, protect and manage historical and cultural sites and natural aesthetic values of marine and estuarine areas, for present and future generations.
- To facilitate the interpretation of marine and estuarine systems for the purposes of conservation, education and tourism.
- To accommodate within appropriate management regimes a broad spectrum of human activities compatible with the primary goal in marine and estuarine settings.
- To provide for research and training, and for monitoring the environmental effects of human activities, including the direct and indirect effects of development and adjacent land-use practices.

Clearly these generic objectives cover the field for national policy but for the development of a plan for a specific area they need to be given a specific local focus. Thus, for example:

- To restore and protect biological communities by excluding all fishing and collecting in a core zone comprising at least 10% of the area.
- During the nesting and hatching season, to prohibit access to an island of special significance as a sea turtle nesting area, other than for purposes of approved research.
- During spawning season, to prohibit access to prime spawning sites of specified commercial species.
- To establish and implement targets for water quality.
- Where practicable, to employ and involve local residents in management.
- To encourage compatible recreation and tourism in appropriate parts of the area, with particular emphasis on training, recreation and tourist leaders.
- To provide continued access to part of the area for limited fishing for pelagic species by traditional inhabitants.
- To define key parameters to indicate usage levels, environmental health and maintenance of viable biological communities and to implement appropriate monitoring.

5.3 TECHNIQUES AVAILABLE TO MANAGERS

A management plan establishes a framework for controlling human activity in order to achieve specified objectives. There is a limited range of techniques that may be applied, singly or in combination, to control the interaction between human activity and the environment. The material here is taken from Kenchington (1990), where it is discussed in more detail.

5.3.1 Prohibition

Absolute prohibition of access to a prescribed area is the simplest form of regulation. It is easily understood and the resulting management task is clear. If a person enters the prescribed area it is clear that an offence has been committed.

Complete prohibition of access for reasons of conservation is rare. Arrangements for access are usually required, at least for research and monitoring even in the most highly protected areas.

The simplicity of prohibition as a control mechanism is that it establishes a black/white or yes/no basis. Management control consists of determining whether or not the prohibited access or activity occurs, followed by enforcement action as appropriate.

The next most simple control is prohibition of a class or classes of activity within a prescribed area. This is widely used in national parks and protected areas to exclude incompatible activities such as hunting, fishing or habitat destruction.

5.3.2 Limitations

In a national park or reserve some activities, particularly those relating to recreation, nature appreciation and research, are usually allowed and often encouraged. The task of management is to protect the structure and processes of the protected area by enforcing measures that limit human activities to levels that can be sustained without causing significant unacceptable impacts.

Limitations may also be an appropriate approach to management where two activities are sustainable but may interfere with each other's amenity. Examples include interactions between sailing vessels and power boats, and between some forms of research and public recreation.

The problem of management by limitations is that it is generally much more complex and demanding than simple prohibition. The management regime is harder for users to understand and for managers to enforce. Six general types of limitation strategy are widely used in marine resource and area management, as follows.

Limitation by spatial control

A regulation may apply to a part or parts of the protected area. At its simplest, spatial control defines the area of application of a management scheme. More elaborately, it can be used to establish zones and thus to assign uses within a managed area.

A zoning plan will subdivide a managed area into two or more sub-areas with defined activities allowed in some areas but not in others. This may be achieved either by specifying the permitted uses and stating that other uses are prohibited or by a presumption in favour of any activities unless specifically prohibited. The difference between these approaches can be very important in the planning and management processes.

Specifying permitted uses has the effect that no activity is allowed unless it is specifically permitted by the manager. The advantage for the protected area manager is that the proponents of a new form of use must demonstrate that it is not inconsistent with the objectives of management before it is permitted.

The alternative is that any activity is allowed unless it is specifically prohibited. This is more in accordance with Common Law freedoms and related concepts of individual liberty but has the disadvantage that the onus of proof is on the managing agency to demonstrate that a new activity is inconsistent before limiting or prohibiting it. To do this may require a long and costly programme of monitoring or research, during which period the use becomes established and may have a substantial impact. The approach of specifying permitted uses is most commonly used for national parks and similarly protected areas.

Limitation by temporal control

The needs and capacity of management are not constant in time. On the shortest time scale there may be a need to prohibit access to an area during part of the 24-hour cycle when it is particularly vulnerable to disturbance or when it is not possible to provide adequate management supervision.

Similarly, many species may be more vulnerable and need higher levels of protection at predictable periods within a year. Such vulnerability generally relates to the reproductive cycle.

On a longer time scale, there may be a need to provide increased protection to an area for a period of one or more years. This may be required as part of a management strategy to allow for recovery after a period of recreational use has led to cumulative impacts, for example trampling on vegetation or diving on corals. It may also be needed to protect an area from human impact while it is recovering from a catastrophe such as a severe storm or pollution event.

Limitation by equipment restriction

Some forms of equipment or techniques of use may be efficient for their purpose in the short term but have damaging impacts which make them inefficient in the long term. This is particularly the case with a number of fishing techniques. Much fisheries regulation involves banning or limiting the efficiency of equipment in order to limit impacts on the stock or its environment.

Non-fishery equipment restrictions may be applied to the use of anchors that damage sea-grass or corals, or to the speed of vessels in sheltered waters where the wake may cause erosion of banks or shorelines.

Limitation by quotas

An alternative to the above options is to set limits so that the harvest does not exceed the capacity of target natural resources or the environment. In some marine protected areas recreational or subsistence fishing may be consistent with overall management objectives. Appropriate limitation may be achieved by applying a quota on a daily basis through a bag limit that restricts the number of fish a person may take in a day. In some fisheries, a quota may be established for a season with the catch being carefully managed and the fishery closing once the quota has been caught.

In other management systems, often applied to recreation or tourism, quotas may take the form of allocation of a predetermined carrying capacity designed to protect the resource. The determination of carrying capacity is a complex process involving scientific assessment of the amount of use a site can sustain without significant degradation. It also involves socio-economic consideration of the nature and extent of the experience or activities that should occur within the ecological threshold of that site. The assessment is complicated by the fact that quite small changes in the way an activity is carried out can cause large changes in the impact of the activity, and hence in the carrying capacity. An example is the reduction in impact of snorkellers when they are trained and required to avoid touching coral.

Recreational or tourist quotas may be applied in a number of ways. Thus the number of people allowed on to a beach or island to carry out an activity, such as diving, may be limited at any one time. Again, the quota may be applied on a daily or seasonal first-come, first-served basis or allocated to individuals. For example, first-come, first-served may be applied through a system whereby moorings are provided and no further boats may enter a site once all the moorings are occupied. Allocated quotas are widely used in the regulation of tourist operations providing security of access to a specific site or group of sites.

Limitation by skill licences

Most fisheries and boating management schemes have some system whereby individuals are required to demonstrate knowledge and technical competence before they are allowed to operate without the supervision of someone experienced in fishing and boat operation. Licensing is becoming increasingly important in marine recreation and tourist activities, such as scuba-diving or speedboat driving, where incompetent operators may cause unnecessary impacts and engender considerable public risk. Skill licensing, particularly if accompanied by the possibility of withdrawal of the licence for noncompliance, can be used to establish codes of practice to minimize adverse impacts of an activity. Knowledge of management regulations and how they apply can be made a compulsory part of a training and evaluation programme for a skill licence. Failure, particularly repeated failure, to abide by regulations can then be grounds for loss of licence.

Resource allocation licences

A skill licence demonstrates competence to undertake a particular activity. Where resources are limited, or where there are social or economic reasons to limit access to a resource, quotas may be issued through the allocation of licences. The difference between a quota and a resource allocation licence is that while the quota specifies the maximum extent to which the holder or an entire class of users may undertake the activity in question, the licence limits the number of people who may undertake the activity but does not necessarily limit the extent to which they undertake it.

Limitation by permit

A permit system provides scope for management discretion in establishing specific conditions under which a potentially damaging or disruptive activity may occur. It can be extremely demanding of management resources, because each application has to be treated on an individual basis and it is generally expected that this will be done quickly. A permit system can provide detailed consideration that is impracticable in general plans.

Most plans contain a mixture of techniques for the regulation of use. They may be used in complex combinations.

5.4 ASSEMBLING A MANAGEMENT PLAN

The planning approach discussed here is developed from that outlined in the UNESCO Coral Reef Management Handbook (Kenchington and Hudson, 1984) which was in turn developed from processes evolved in management planning for the Great Barrier Reef Marine Park (Kelleher and Kenchington, 1982).

Clearly, if the planning programme has to prepare a management plan for an area which has already been designated a Strict Nature Reserve or a National Park, potential uses and the objectives of public consultation are more limited than for zoning an area designated for multiple use.

The actual format selected may range from a small-scale locally adopted municipal plan, such as those developed in the Philippines by Alcala and White (1984), to the nationally endorsed legal instruments required under Australia's Great Barrier Reef Marine Park Act (Kelleher and Kenchington, 1982; Kelleher, 1986; Kenchington, 1990).

Whatever the plan format, it will be necessary during the course of development of plans and associated reports to cover all or most of the following issues:

- Executive Summary – covering the essential issues and necessary decisions. Many of the final decision makers will not have time to read and digest supporting detail.
- Introduction – defining the purpose and scope of the plan and explaining the legislative basis and authority for plan development.
- Statement of the goals and objectives for the planned area as a whole.
- Definition of the area – a formal statement of the boundaries of the planned area and a geographical description of its setting and accessibility.
- Description of the resources of the area – a summary of information directly relevant to decisions should be included in the report. Detail should be restricted to an appendix or separate document.

- Description of uses of the area – this should concentrate on present uses but should place these in the context of past types and levels of use in the absence of a plan. The description should include social and economic analyses of use.
- Description of the existing legal and management framework for activities such as coastal fisheries, marine transportation and other relevant legal controls on present use of the area. Where they still exist or can be recalled, traditional practices of management, ownership or rights to the use of marine resources should be described.
- Analysis of constraints and opportunities for activities possible within the area.
- Statement of the principal threats to the conservation, management and maintenance of the area.
- Statement of inter-agency agreements made or necessary for conservation and management of the area. This may usefully include a summary of consultative processes followed in plan development.
- Statement of the boundaries, objectives and conditions of use and entry for any component zones of the planned area.
- Provision for regulations required to achieve and implement boundaries and conditions of use and entry.
- Assessment of the arrangements (including financial, human and physical resources) required to establish the MPA and to manage it effectively. This should cover:
  - staffing
  - equipment and facilities
  - training
  - financial budget
  - interpretation and education
  - monitoring and research
  - restoration
  - surveillance
  - enforcement
  - evaluation and review of effectiveness.

It is likely that the management plan will have to be supplemented by separate reports and procedure manuals, which will be developed during the life of the plan. Examples of the content of MPA management plans may be found in Salm and Clark (1984) and Kelleher and Kenchington (1992).

The minimum requirements of the planning programme will usually be determined by the legislation providing for management plan development. The programme need not necessarily be restricted to the minimum and generally it will have five phases:

- Initial or premanagement information gathering and preparation.
- Public participation or consultation prior to the preparation of a plan.
- Preparation of draft plan.
- Public participation or consultation to review the draft plan.
- Plan finalization.

The extent and most appropriate means of conduct of each phase will depend strongly on the social and political system under which the plan is prepared.

5.4.1 Initial or premanagement information gathering and preparation

The planning agency, often with the assistance of consultants and research institutions, assembles and reviews information on the nature and use of the area. In many systems this is required in order to prepare a document for declaration or proclamation of a new protected area. Whether or not this has been the case, such information is needed to develop reports and summaries for user and public participation, interagency consultation and consultation with appropriate representatives or experts.

Investigations at the premanagement stage should address three objectives:

1. To establish what is known of the area based on information already available in scientific publications, maps and government reports.
2. To establish the key issues to be addressed in protecting or managing the resources in that area.
3. To determine the need for further information about resources, use and impacts needed for the design of practical management measures.

Kenchington (1984) listed the following questions that should be explored in establishing and preparing a management plan for a marine or estuarine protected area:

- What maps, photographs or satellite images exist of the likely management area?
- What area needs to be physically surveyed to establish boundaries for a management scheme?
- What is known of tides, tidal and other currents in the area?
- What water movement information is available to assist understanding of linkages within and beyond the area?
- What is the geological nature of the sea-bed and coastal land in the area?
- What is the biological nature of the area?
  - community descriptions
  - species diversity
  - dominant biota

  - commercially important species
  - endangered or threatened species
  - changes with time.
- What is the cultural/historic significance of the area?
  - traditional use and impact
  - traditional understanding
  - significance of the area to the culture of local or distant human communities
  - 'stewardship' priorities, for example, caring for the heritage of future generations
  - recreational significance
  - aesthetic significance
- What is the international and regional significance of the area?
- What is the current use and impact?
  - subsistence
  - traditional
  - commercial
  - recreational
  - research
  - tourism
  - shipping
  - dumping and waste disposal
  - alienation, reclamation
  - mariculture.
- Who does what?
- What is the socioeconomic importance of current use?
- What changes to use, impact or environmental condition are likely to occur in the next 5, 10 or 20 years?
- What examples are there of management measures to address similar problems in comparable situations?
  - limited access
  - pollution controls
  - impact prediction and control requirements for new or changed activities
  - education
  - restoration or repair.

The development, testing and refinement of goals and objectives is a primary function of the first phase of public participation and consultation. On the basis of the initial management information, the planning agency, often with the assistance of consultants and research experts, assembles reports and designs other materials for use in the public participation and consultation process.

There are several component purposes that should influence the design of the materials and programme for initial public participation and consultation. These include:

- Presenting information concerning the issues, benefits of management and costs of failure to manage.
- Presenting information summaries prepared by the planning agency concerning the nature and use of the resources of the area.
- Encouraging members of the community to discuss the issues and develop views on desirable future use and management.
- Seeking information on matters and approaches which the community considers should be incorporated in a plan for the area.

The design of the materials to present new and often complex information and concepts in a manner acceptable and comprehensible to users who may have operated for years or generations in the area frequently involves a major educational task. Fishermen and mariners, used to dynamic changes in the state of the sea surface and in the availability of fish or other resources, may find it difficult to accept the concept that human actions can influence the nature of the biological communities of coastal waters and the open ocean. The concept that there are limits to the resources of the sea is a marked departure from the fundamental beliefs and traditions of many seafaring communities.

The design of materials and questions to obtain information without biasing the result is a similarly complex task, particularly for remote communities which may not be exposed to major economic activity, and which often have different cultural and linguistic traditions to those who are seeking information. The fundamental questions may be clear but, to obtain valid information, the appropriate method of asking them involves substantial understanding of the community, its cultural traditions and use of language. Without this, the result may be a reflection of an intent either to present what the questioner wants to hear, or worse, to mislead the questioner.

The equally important design task is to arrange a programme to ensure that the materials reach the intended targets and to establish a timetable for consultations, public meetings and other activities. The programme should be designed to ensure that nobody who has interests in the area to be managed can escape being aware of the planning process, and of opportunities to learn more about the process and to comment on and contribute to it.

5.4.2 Public participation or consultation prior to the preparation of a plan

The agency provides information on the nature and purpose of the planning process. It seeks public and user comment on the accuracy and adequacy of review materials and suggestions for the goals and contents of a plan for the area.

There are two major purposes of the first phase:

1. To present information on the resources and use of the area, on the need for management, and on the objectives and processes of the planning process.
2. To obtain information, to extend and update that presented on resources and uses, and to find out whether the goals and objectives are adequate in addressing community expectations.

It is not possible to propose a 'turn-key' model which would be appropriate, unmodified, in any country or situation other than that for which it is designed.

As an example, the processes of public participation for the Great Barrier Reef Marine Park, described by Kenchington (1990) and Kelleher and Kenchington (1992), are clearly based on a context of high literacy and ready access to print and electronic methods of information distribution. Obviously, different techniques will be needed in, for example, an island nation with a structured village social organization, a strong oral tradition and little local access to print and electronic methods. Here the approach of user involvement should be adapted to convey information and seek information and opinion through socially appropriate techniques, which might include drama, song, storytelling or church meetings (Hudson, 1984).

The essential points are that the usage patterns, expectations, attitudes and local knowledge of users should be determined in the planning stage, and that planning should not be allowed to become the task of remote experts with no direct contact with or understanding of local issues.

There should be two parallel programmes: one of interagency consultation to establish the views of relevant government departments, and one of user or local consultation to gather information which can be fed into the planning process. Consultants may be used for these tasks but it is important that senior officers of the planning agency are sufficiently closely involved to be able to appreciate and evaluate the positions of key user groups with interests in the area.

It is often necessary to develop materials for a number of target groups. It is desirable that each user group becomes aware of the issues and concerns important to the others. A single information package has advantages but is often impractical because different groups may require information in different forms.

The groups to be contacted should be identified early in the planning process (Gawel, 1984). They are likely to include:

- Local resource users.
- Researchers and technical specialists.

- Industry lobby groups.
- Conservationists.
- Non-government organizations.
- Community opinion leaders.
- Local government.
- Other government agencies.

The programmes should be developed with people who know and understand the lives and values of people in the communities. In an open democracy it may be reasonable to expect that individuals will feel free to comment to government officials or consultants. Nevertheless, it is possible that elected officials or self-appointed community representatives will convey information and views that do not take account of majority interests, possibly leading to strong distrust and cynicism regarding the process and extreme reluctance to offer any information or opinion. In some situations, and particularly in communities little affected by economic development, there is a strong hierarchical structure with hereditary or appointed leaders who speak for the community. Here, approaches to other community members for additional information and opinion will generally fail and, as a breach of accepted protocol, they will cause embarrassment and distrust.

In practice, a combination of techniques is probably needed to achieve an understanding of the importance of a marine area to the communities that use it, and of the views of people more distant from the area but concerned with its management.

5.4.3 Preparation of draft plan

The agency, possibly with the assistance of consultants, prepares a draft plan consistent with the legislation, taking account of information and views obtained in the preceding phase. Specific objectives should be defined for the plan as a whole and for any component zones or sub-areas within the plan. The agency should also develop explanatory materials for the next phase.

As with the initial phase, materials should be designed to achieve simple and effective communication with the range of user and interest groups. Similarly, a timetable of meetings, consultations and other activities should be established as a framework for public discussions on the draft plan.

The components that should influence the design of the materials and program for public participation and consultation to review a draft plan include:

- Summarizing the results of the first phase of public participation with particular emphasis on presenting the expressed views and priorities of user or interest groups.

- Presenting updated information reflecting new knowledge of resources and usage resulting from the first phase or since obtained.
- Explaining the goal and objectives of management.
- Listing the guidelines used in plan design.
- Stating briefly the reasons for the selection of areas with substantial restrictions of activity.
- Encouraging members of the community to discuss the issues and present views on the adequacy of the goals and objectives of management, and of the plan in meeting those goals.
- Seeking specific reaction to alternative solutions presented in the plan or accompanying materials.
- Seeking alternative solutions that are consistent with the goals and objectives but that have less impact on the amenity of users.

5.4.4 The second phase of public participation

The agency provides information on the results of the initial phase, explains the design and concepts of the draft plan and seeks public and user comments on the content of the draft plan. The process is similar to the initial phase, but the purpose of the programme is more easily explained to the public. Whereas the first phase explained the concept and sought information and management suggestions, the second seeks reactions to a plan containing specific proposals. Individuals and interest groups can study those proposals and consider the degree to which they would be affected or to which their concerns are addressed.

In general the approach and the range of individuals and agencies consulted should be similar to those for the first phase of public participation and consultation.

The processes of analysis and reporting will be similar to those for the initial phase although the task may differ slightly, since participants are reacting to specific proposals rather than providing information and general suggestions. The strength and degree of acceptance of views or reactions to specific aspects of the draft plan may be critically important factors in decisions on changes to be made in finalizing the plan. The analysis should reflect the strengths and limitations of the information gathering technique in terms of representation of views of the community or specific sectors of the community.

5.4.5 Plan finalization

The agency finalizes the plan, taking account of comments received on the draft, and submits it for approval and adoption as specified in the legislation or in accordance with well established or consistent local practice.

Analysis of the public review of the proposed plan will reveal the degree of general acceptability, which should be reasonably high if the plan reflected the views and information collected in the initial phase of public participation. It will usually reveal some generic dissatisfaction from one or more groups that consider themselves disadvantaged or ignored in the draft. It will also shed light on specific situations where the solution proposed in the draft plan should be revised. Such situations arise where new information emerges because the proposed solution would prevent a use or activity that was not previously declared or identified. They also arise where the draft plan contained a marginal decision or an optional choice of site for particular activities.

Revision of the draft plan to produce a final version for official adoption is again a matter for some judgement in the light of community attitudes and expectations of participatory decision-making. If there is no change between the draft and the final plan, the community is likely to view the consultative process as insincere. Conversely, if there are numerous changes there may be demand for a further phase of consultation because the final plan does not resemble that which was commented upon. Changes between draft and final plan will generally address problems identified in the draft by one user or interest group but, unless care is taken, they may create problems for another by changing a situation that was seen as satisfactory. Some informal consultation with representatives of key user groups may be needed in order to ensure that the solution to one problem does not result in the creation of a greater problem.

The formal adoption of the measures to be incorporated in the final plan is usually a process covered by legal and constitutional procedures. The result is usually one or more documents written in formal legal language that describe the boundaries of the managed area, and any component subareas within it, and set out the regulations required to give effect to the conditions of use or entry developed in the plan.

5.5 PREPARING FOR MANAGEMENT

The management plan should have been developed with a clear view of the practicalities of implementation and thus of the resources needed, and likely to be available for management. Nevertheless, it is often the case that the implementation of the plan will be the task of a different group of people from that which developed it. The implementing group will need to develop materials for staff training and public education on the provisions of the new plan. It is therefore important that, in addition to the formal finalized plan, the planning process should provide the management team with adequate material to explain the plan, the key decisions and the major outcomes of the public participation processes.

Ideally there will be more substantial handover with one or more members of the planning team following through to provide continuity in the establishment process.

The overall management programme should generally have seven key elements:

- Training – to ensure that those who implement the plan understand it, their duties under it and the management approach they should use.
- Education – to ensure that the public and users of the managed area are aware of the existence of the plan and their rights and responsibilities under it.
- Surveillance or activity monitoring – to assess how people are using the area and the extent of compliance with the plan.
- Enforcement – to follow up infringements of the plan and, by education or legal action, to seek to ensure that they are not repeated.
- Monitoring – to provide up-to-date awareness of the condition of the managed area and of the impacts upon it from human use and other factors.
- Impact prediction and management – to assess in advance the likely impact of significant new or altered uses proposed for the area, to determine whether such uses can be consistent with management objectives for the area and, if they are so, to establish conditions under which such uses may be allowed.
- Review – to revise and improve the plan after it has been in operation for a defined period in order to take into account the results of surveillance, monitoring, research, increased understanding of the nature of the area and the impact of use, and changes in public attitude to and perception of the area.

It takes some time to establish management. It is therefore important that there is an adequate initial establishment phase with continuing support to enable management to become firmly established.

Commitment to continued implementation is often the weak link of the protected area process. The innovation and drama of proclaiming a protected area and developing a plan are attractive and lead relatively easily to investigation and planning commitments by governments and international agencies. To have any lasting effect, creation of a protected area entails a perpetual commitment of resources of finance, equipment and personnel to routine management (Machlis and Tichnell, 1985). This is not easily sustained in competition with other calls on government resources.

The ideal solution is one that leads to management being self-funding from revenues generated as a consequence of the operation of the protected area. This ideal requires a high standard of management since if any users are required to pay they will expect to see evidence of the management they are supporting.

REFERENCES

Alcala, A.C. and White, A.T. (1984) Options for Management, in *Coral Reef Management Handbook*, (eds R.A. Kenchington and B. Hudson), UNESCO, Jakarta, pp. 31–40.

Australian Committee for the International Union for the Conservation of Nature and Natural Resources (1986) *Australia's Marine and Estuarine Areas – A Policy for Protection*, ACIUCN Occasional Paper 1, 30 pp.

Gawel, M. (1984) Involving Users in Management Planning, in *Coral Reef Management Handbook*, (eds R.A. Kenchington and B. Hudson), UNCESCO, Jakarta, pp. 99–109.

Hudson, B. (1984) User and Public Education, in *Coral Reef Management Handbook* (eds R.A. Kenchington and B. Hudson), UNESCO, Jakarta, pp. 147–160.

Kelleher, G. (1986) Managing the Great Barrier Reef. *Oceanus*, **29**(2), 13–19.

Kelleher, G. and Kenchington, R.A. (1982) Australia's Great Barrier Reef Marine Park: Making Development Compatible with Conservation. *Ambio*, **11**(5), 262–269. Reprinted (1984) in *National Parks, Conservation and Development: the role of Protected Areas in Sustaining Society*, (eds. J.A. McNeely and K.R. Miller), Smithsonian Institute Press, Washington.

Kelleher, G. and Kenchington, R.A. (1992) *Guidelines for Establishing Marine Protected Areas. A Marine Conservation and Development Report*, IUCN, Gland, Switzerland, vii + 79 pp.

Kenchington, R.A. (1984) Making a Plan, in *Coral Reef Management Handbook*, (eds R.A. Kenchington and B. Hudson), UNESCO, Jakarta, pp. 115–121.

Kenchington R.A. (1990) *Managing Marine Environments*, Taylor and Francis, New York, 250 pp.

Kenchington, R.A. and Hudson, B. (eds) (1984) *Coral Reef Management Handbook*, Jakarta, UNESCO, xx + 281 pp.

Machlis, G.E. and Tichnell, D.L. (1985) *The State of the World's Parks: an International Assessment for Resource Management, Policy, and Research*, Westview Press, Boulder, Colorado, 131 pp.

Salm, R.V. and Clark, J.R. (1984) *Marine and Coastal Protected Areas: A Guide for Planners and Managers*, IUCN, Gland, Switzerland, 302 pp.

6

Techniques for managing marine protected areas: zoning

Dan Laffoley

6.1 INTRODUCTION

The need for and difficulties of managing uses of the sea so that they are ecologically sustainable forever are common to all coastal waters (Kelleher, 1993). The problems related to the development of effective sea management is something that many countries have been struggling with for the last couple of decades. Such efforts have increased substantially in recent years, reflecting the growing realization that many of our current practices are unsustainable, and as clearer evidence occurs of growing damage to coastal ecosystems.

The concept of integrated planning and management of coastal waters is increasingly seen as holding many of the potential solutions to existing problems. One of the challenges, however, of integrated planning and management of coastal waters in general, and marine protected areas (MPAs) in particular, is developing a high level of understanding and compliance in the people who use such areas. This is fundamental to the success of any management regimes. If coastal populations who use the sea and its resources do not understand, value and support the aims of any management project, it is unlikely that the scheme will succeed.

One of the most effective ways in which support and understanding have been achieved is through user-related zoning schemes. This chapter examines the principles behind the development of effective zoning schemes and shows how this approach is central to the management process and quite often the main reason for its success. Four case studies are used to illustrate this approach. These are ordered chronologically, and by decreasing area, starting with the Great Barrier Reef Marine Park in Australia and concluding with the zoning of Lundy Island Marine Nature Reserve in the UK.

Marine Protected Areas: principles and techniques for management. Edited by Susan Gubbay.
Published in 1995 by Chapman & Hall, London. ISBN 0 412 59450 1.

6.2 ZONING SCHEMES

The IUCN (the World Conservation Union) has derived a primary goal for marine management (Kelleher and Kenchington, 1992):

> To provide for the protection, restoration, wise use, understanding and enjoyment of the marine heritage of the world in perpetuity through the creation of a global, representative system of marine protected areas and through the management in accordance with the principles of the World Conservation Strategy of human activities that use or affect the marine environment.

Achieving this aim is particularly difficult due to the inherently linked nature of the sea. Quite often protection or control of activities in a small area becomes meaningless if the influence of activities outside the site (e.g. water quality) is more significant. There is therefore a need to ensure that in marine management planning core areas are protected by sufficiently extensive buffer zones and also to apply management to large enough areas. The concept of applying management to 'large marine ecosystems' such as the Yellow Sea, the Faroe Plateau and the Humbolt Current ecosystem is now being advocated as 'a useful means for addressing the many-faceted problems of sustainable development of the world's oceans' (Sherman and Laughlin, 1992).

MPAs generally cover smaller areas, although the Great Barrier Reef Marine Park is one exception. For such sites the IUCN goal for marine management can be interpreted as being concerned with ensuring that a particular marine protected area can sustain, in the long term, existing populations of wildlife and natural features whilst recognizing cultural values and continuing sustainable exploitation of marine resources. Such aims are delivered through management plans and zoning is often an important part of such a plan.

The development and maintenance of a successful zoning scheme to manage activities in an MPA is a complex process which needs to take account of a wide range of variables which change according to time. These include:

- knowledge of the physical and biological characteristics of the area;
- knowledge of user activities, resource use and perceptions;
- knowledge of conflicts between different users;
- knowledge of conflict between users and the environment.

Traditional management of marine areas has employed a number of management strategies including:

- complete closure of areas;
- periodic closure of areas;

- regulation of effort;
- equipment constraints;
- impact limitations;
- habitat restoration.

Although each strategy is useful, either individually or together, they do not represent a comprehensive management plan (Hopley, 1989). The individual strategies also have an inherent weakness in that they are largely technical and are highly specific measures which do not relate well to general users of the area. Quite often users are almost totally unaware of many regulations for an area. They often do not read, let alone understand, technical marine legislation and seldom have the interest or inclination to read large management plans. This can be seen as a failure of the relevant authorities to explain the issues with sufficient clarity or present it in an appropriate form.

Zoning schemes, however, provide the level of integration required in an easily understandable format to overcome many of the problems exhibited by less comprehensive single-interest approaches. They do not replace management plans, which are essential documents to guide the actions of regulatory and management bodies surrounding or involved in the MPA. They do, however, provide simple mechanisms to explain the management process to a wide public audience in an understandable format.

Kelleher and Kenchington (1992) define a zoning scheme as the means by which planners and managers define the purposes for which a protected area may be used. Although a zoning scheme may be a legal document, it must be understood by those whose actions it seeks to control and a key to this lies in effective and culturally appropriate involvement or consultation of users in the process of plan development.

Activity-based zoning is realistic because the management of MPAs is predominantly about managing how people use or do not use an area. It is rarely about managing the resource, e.g. planting mangroves. The focus on activities can also help users to feel that their interest in the area is being considered. They will be asked about their 'use' in the formulation of the zoning scheme and therefore it is very obvious that it is being taken into account in the management proposals.

A negative aspect of zoning is that it can seem a very rigid process which is too simplified in a complex world. People may comment that it is not easy or appropriate to draw boundaries in the sea but this can be addressed by pointing out that it is the activities that are generally being controlled, and these can be limited to clearly defined boundaries. In addition, as zoning schemes are revised on an appropriate time-scale, the zones are very unlikely to be set in stone.

In some situations zoning schemes can affect what happens outside MPAs. If, for example, a successful scheme finds public support, it may lead to calls for wider application. Another effect may be to increase and concentrate activity around the boundaries of a zone, such as a no-fishing area. Managers should accordingly be alert to the wider implications of such zoning schemes.

6.3 THE BASIS FOR ZONING

A key consideration in the preparation of a zoning scheme is the basis on which zoning will be developed. Options can range from a purist approach of zoning based entirely on the ecological sensitivity of habitats, through to zoning related purely to human activities. In reality zoning schemes fall somewhere between these two extremes, taking account of both habitat sensitivity and user requirements.

Perhaps falling nearest to the purist ecological approach is the zoning of **biosphere reserves.** Biosphere reserves are protected areas of representative terrestrial and coastal environments which have been internationally recognized under the UNESCO MAB (Man and the Biosphere) Programme for their value in conservation and in providing the scientific knowledge, skills and human values to support sustainable development. Biosphere reserves are united to form a world-wide network which facilitates sharing of information relevant to the conservation and management of natural and managed ecosystems (UNESCO, 1987).

To understand the role zoning takes in the management of biosphere reserves it is necessary to understand their functions. For a site to be a true biosphere reserve it must display three basic functions which are usually presented in three complementary roles (Batisse, 1990):

- the conservation role (conservation of genetic materials and ecosystems);
- the logistic role (participation in the international research and monitoring network);
- the development role (cooperation with local populations in or around the biosphere reserve to promote forms of sustainable development compatible with the conservation objectives).

Zoning is used within biosphere reserves as the essential way of combining all three roles with **core areas**, which have to be strictly protected to meet the conservation objectives: the **buffer zone**, which should be clearly delimited for management purposes; and the **transition area**, which can extend over the territory where cooperation with local people for sustainable rural development can be organized. In these situations the zoning and delineation of core areas is closely related to ecological sensitivity but with an acceptance of the role people have in the protected area (Figure 6.1).

The objectives of the biosphere reserve scheme are clearly appropriate to marine environments but the guidelines developed for terrestrial biosphere

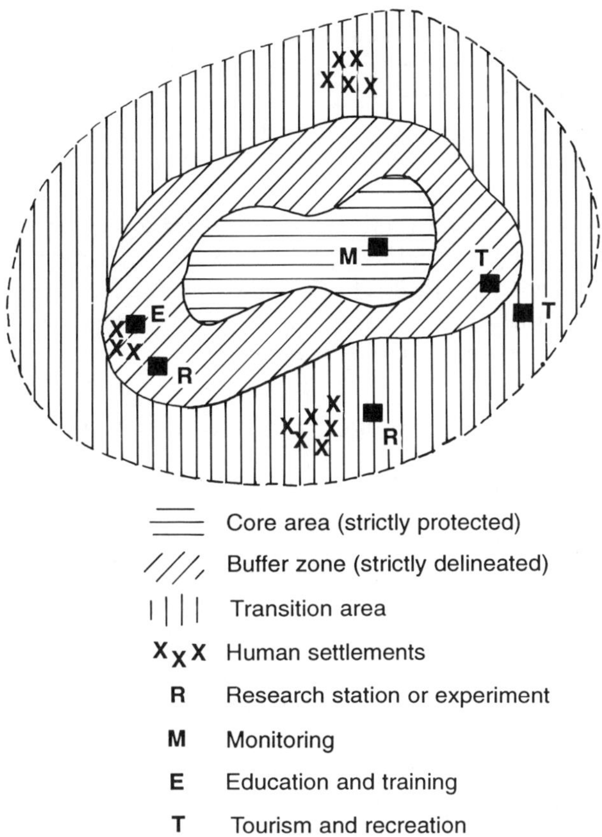

Figure 6.1 Schematic zonation of a biosphere reserve. The core area is devoted to strict protection. The transition area is not necessarily delimited, but implies involvement of local population.

reserves will require modifications if the approach is to be successfully applied to the management of coastal and marine areas (Kelleher and Kenchington, 1992). In particular the scale of many marine ecosystems, and the nature and scale of the processes of fish population recruitment and maintenance, are issues that have not, so far, been satisfactorily tackled. Examples where the biosphere reserve concept is being tried in the marine environment include the Central California Coast Biosphere Reserve, the Galapagos Islands Reserve and the Mer d'Iroisse Biosphere Reserve in France. These and other examples are described in Price and Humphrey (1993).

A more user-related approach is that developed by the Great Barrier Reef Marine Park Authority in Australia. Faced with the legislative requirement to place 344 000 km^2 of the Barrier Reef under active management, they have used multiple-use zoning schemes, portrayed on navigation-quality maps, as the principle way to achieve this. Such zoning plans have been drawn up through multiple stages of public consultation and allow for reasonable use of the reef for recreation and commerce, whilst at the same time ensuring that areas are set aside for strict protection. Such an approach is not wholly determined by habitat sensitivity and the location of many zones with a higher degree of use is clearly linked to managing human activities themselves rather than the reef resource. This approach is considered in more detail as a case study later in this chapter.

6.4 THE FORMAT OF A ZONING SCHEME

An increasing number of zoning schemes tend to conform to a format which depicts use-related management information as overlays on marine charts using colours portraying the levels of habitat protection (Plate 1). The most effective of these pays close attention, either consciously or unconsciously, to the way people are conditioned to react to colours. At the simplest level shades of orange, pink and red represent areas of highest habitat protection whilst areas set aside for recreational purposes are represented by shades of green. We all recognize red as a warning or 'no-go' signal, while green indicatess safe or 'go'. In wider areas of sea surrounding restricted areas, shades of blue are often used to represent transition or buffer zones where most existing sea uses can continue.

The format of the zoning scheme should remain simple. There is a temptation to use a large range of colours and a scatter of very small zoning areas within a comparatively small area of sea, each reflecting some subtle variation in restriction. This usually becomes too complex to be implemented effectively. Experience show that it is better to design a simpler zoning scheme, understandable to all and simple enough to be policed effectively. If it proves impossibly complex to translate management into a zoning scheme, then it is highly probable that such a management approach would fail due to the inherent weaknesses related to its complexity.

An essential part of any zoning scheme is the table that relates activities to coloured areas on the chart. Usually this lists major categories of activities down the left-hand side and coloured zoning areas across the top, and indicates where activities can occur by using 'yes', 'no' or 'limited' in the appropriate box. Experience of existing zoning schemes suggest that showing the lowest level of restriction on the left of the table, with many boxes beside activities indicating 'yes', creates a more positive impression to users of an area and balances the often negative impression created by the designation of the MPA. Additional information relating to navigation rights and further guidance on activities indicated as 'limited' can be provided underneath as footnotes.

Zoning schemes which deviate from this approach can be very misleading on first impressions. Existing zoning schemes which clearly do not fulfil their potential include one where restrictions are represented by a shade of green and recreational areas by a shade of orange and another where crosses have been used on the right-hand side of the zoning box to indicate where activities can occur, with all other boxes being left empty. Adopting the correct format for the zoning is of ultimate importance to the success of the MPA management process.

6.5 OBJECTIVES OF ZONING

The following section is based on the guidelines produced by Kelleher and Kenchington (1992) for the development of zoning schemes. The principle objectives of zoning reflect the objectives for the management of the MPA and are usually:

- to provide protection for critical or representative habitats, ecosystems and ecological processes;
- to separate conflicting human activities;
- to protect the natural and/or cultural qualities of the MPA whilst allowing a spectrum of reasonable human uses;
- to reserve suitable areas for particular human uses, whilst minimizing the effects of those uses on the MPA; and
- to preserve some areas of the MPA in their natural state undisturbed by humans except for the purposes of scientific research or education.

Given the variability of the marine environment and socio-economic factors it is not possible to propose a single model which will result in the development of effective zoning schemes for different MPAs. Examination of zoning schemes from MPAs around the world, however, indicates some other important aspects which need to be taken into account in developing future schemes (Hopley, 1989):

- Large management areas are easier to zone than smaller areas.
- Zones should encompass all acceptable uses.
- There should be a gradation from least through to most heavily protected zones to buffer protected areas.
- Careful use of colour ensures 'automatic' understanding of the level of protection and regulation within the different zones.
- Traditional users of the managed area should be consulted and involved in the development and implementation of management plans.

- Traditional law and management practices consistent with the goals and objectives of modern management should be incorporated to the greatest possible extent.

6.6 DEVELOPMENT AND IMPLEMENTATION OF ZONING SCHEMES

The development and implementation of an effective zoning scheme, as with marine management in a more general sense, probably relies more on effective interaction with the public and users of an area than with any other single aspect of the site management process. The science behind the process must be sufficiently robust to support the decision-making process but repeated experiences around the world have shown that only the schemes which directly relate to the users of the area tend to succeed. Laffoley (1994) examined the reasons for the apparent success of a number of MPAs in Australia and New Zealand and defined a number of guiding principles (Laffoley *et al.*, 1994). These management principles are as follows:

- Achieve a balance between the 'top down' legislative agency and the 'bottom up' community involvement approaches to site management. The idea is to strive for conservation through collaboration achieved by local discussions throughout the MPA management process.
- Make a strong link with the users of the area throughout the implementation of site management by preparing it from a user's viewpoint. The most effective mechanisms developed so far involve zoning schemes where the management proposals and subsequent outcomes are depicted on navigation-quality charts, using colours to portray zones where activities can occur and to show the differing levels of management restrictions. A simple coloured map can be very effective at getting this measure across. This approach highlights the importance of using easily recognized boundaries to the MPA and internal management zones (ideally straight lines joining significant, recognizable features).
- Manage human activities in a structured way within the MPA boundaries. This principle relates closely to the previous one. The success of site management depends on a clear user-related framework.
- Develop management of the MPA-related marine resources based on long-term resource protection and sustainable utilization. Twenty-five years has been the most frequently used planning time-scale. This long-term view moves the emphasis from current conflicts to what individuals or organizations wish to see in the future. This makes the process much more positive.
- Make the aims and goals within the site boundary clear from the earliest stages. It is vitally important that this is established clearly and briefly

from the earliest stages of implementation of site management. The absence of such guidance will lead to individual interpretations and runs the risk of longer-term problems in establishing effective site management.

- Understand that marine management is a dynamic process. Marine management must be a dynamic process reflecting changes of activity or increased knowledge of marine resources, wildlife and natural features. Periodic revisions of zoning schemes and underlying management plans are essential to the maintenance of effective management regimes.

At a more detailed level, and in relation to the practical process by which zoning can be developed, Kelleher and Kenchington (1992) suggest from experiences with existing zoning schemes that there are ideally five stages to the process. These are described in detail in Chapter 5. Two additional stages are also important and contribute towards the success of any zoning scheme: dissemination of information, and on-site zone marking.

Dissemination of information

A vital part of the process is the production of user-related information and effective dissemination. Coloured zoning schemes can be supplemented by brief leaflets tailored to the needs of individual user groups, as have been produced for sections of the Barrier Reef Marine Park. In other situations the careful use of television to promote the consultation process, using community service announcements, or the use of video to explain processes and procedures to user groups, such as fishermen, have been very successful. There is, however, no real substitute for face-to-face meetings with the public and communities who live around MPAs. Meetings to disseminate information on a first-hand basis should form an integral part of the implementation process and should occur as early as possible.

Marking zones 'on site'

To accompany the development of zoning schemes it is important to attempt to mark significant boundaries 'on site'. Key boundaries are usually considered to be the outer boundary of the MPA, within which structured management has been implemented, and the boundaries of core areas, which may be the principle non-extractive area of the MPA.

To be effective, attention needs to be given to boundary marking in the development of a zoning plan. Straight lines, linking significant landscape features, should be used when possible as curved lines are difficult to visualize at sea and are hard to enforce in any meaningful manner. The exception to this is for island systems when a distance from the shorelines can be used to some effect.

The marking of boundaries can take several forms. For land-based systems the approach adopted in New Zealand for marine nature reserves provides a useful model. The lateral boundaries of some reserves are marked using white triangular signs on the shore, which when aligned indicate the extent of the reserve. This approach is similar to the use of signs for channel markers to enter harbours and therefore can be used easily by seafarers. For water-based site marking, some countries use systems of buoys (e.g. the Looe Key National Marine Sanctuary in the United States). This approach is limited on cost grounds to relatively simple zoning schemes and only works in areas with slow tidal streams.

6.7 CASE STUDIES

The case studies in this section illustrate the flexibility of zoning schemes and how they can be applied in a range of locations, and over a wide span of geographical scales.

6.7.1. The Great Barrier Reef Marine Park

The Great Barrier Reef was probably one of the first places in the world where large-scale zoning was developed as an integral part of the management process.

The Great Barrier Reef Marine Park is a multiple-use protected area. It meets the criteria for selection and management as a biosphere reserve, although it has not been formally proposed or established as one. The Reef was inscribed on the World Heritage List in 1981 as a natural site.

Through the use of zoning, conflicting activities are separated. Areas are provided which are suitable for particular activities, and some areas are protected from use. Levels of protection within the Park vary from almost complete absence of restriction on activity to zones within which almost no human activities are permitted. The only activities which are prohibited throughout the Park are oil exploration, mining (other than for approved research purposes), littering, spear-fishing with scuba, and the taking of large specimens of certain species of fish (Kelleher, 1993) (Plate 2).

To zone the reef the Marine Park was divided into four main sections and individual zoning schemes were prepared for each section after several stages of public consultation. In the schemes which have been developed so far, there are three major categories of zone:

1. **Preservation zones**. Equivalent to IUCN Category I (Scientific Research/ Strict Nature Reserve). The only human activity permitted is strictly controlled scientific research.
2. **Marine National Park zones**. Equivalent to IUCN Category II (National

Park). The major uses permitted are scientific, educational and recreational.

3. **General use zones.** Equivalent to IUCN Categories IV (Managed Nature Reserves) and VI (Resource Reserves). Uses are held at levels which do not jeopardize the ecosystem or its major elements. Commercial and recreational fishing are generally permitted, although bottom trawling is prohibited in one of these two zones.

The zones are fixed during the life of a zoning scheme (usually 5–7 years). They are complemented by:

- generally smaller areas which give special protection from time to time to animal breeding or nesting sites;
- sites in General use;
- other zones protected to allow appreciation of nature free from fishing or collecting; and
- sites suitable for scientific research.

Because there has been a dramatic increase in the use of the Marine Park by tourists, the original zoning system (which focused on fishing) proved inadequate. There is increasing competition for tourism use of some sites. Usually these are near major areas of coastal development (e.g. Cairns or Townsville), or have particular attributes which make them suitable for tourism.

At a special conference arranged by the Marine Park Authority in late 1988, participants agreed that there was a need to incorporate a tourism strategy into the zoning system. This identifies areas that are particularly suited to tourism development and those that should be retained in their natural state, undisturbed by such development.

The strategy is being implemented through the zoning system, but it was included in the re-zoning of the Cairns Section of the Marine Park. It will be extended to the other sections as they are re-zoned during the next few years. Initial zoning of the entire Marine Park was completed on schedule in 1988.

A major constraint in zoning for tourism has been that many tourists like to observe the natural qualities of the Great Barrier Reef undisturbed by fishing. Modern technology allows them to do so from semi-submersible vessels and underwater observatories. If carried out with care these activities have very little, if any, effect on the Reef's ecosystem. The need to provide for such activities in zones protected from fishing has led the Authority to divide all zones, with the exception of the Preservation Zone, into two categories:

- Category 1 – no structures (for example, floating hotels, pontoons or mariculture) will be permitted. Mooring buoys may be permitted.

- Category 2 – such structures are permitted provided they meet environmental guidelines. A permit is required.

The adoption of this new zoning scheme allows the Authority, in association with interested members of the public and with other agencies, to develop and apply a tourism strategy for the whole of the Great Barrier Reef Marine Park. The aim is to ensure that the Reef will not become dotted with tourist and other structures but, at the same time, to allow for careful development on reefs which are suitable for that purpose. This strategy should allow the Authority to provide for ecologically sustainable development.

6.7.2. Solitary Islands Marine Reserve

The Solitary Islands Marine Reserve in northern New South Wales, Australia, is a multiple-use reserve which uses the zoning principles established for the Great Barrier Reef Marine Park.

The reserve covers about 75 km of coastline directly north of the city of Coff's Harbour, an area of predominantly rocky headlands and sandy bays interspersed with small estuaries and a number of offshore rocks and islets. The marine life of the area is outstanding – including a wide variety of corals supported by warm currents from further north and, reputedly, the densest population of clown fish and anemones in the world.

The idea of creating a marine reserve came from the local community in 1969, in response to the threat of having an international spear-fishing competition in the area. The community asked New South Wales Fisheries to protect the area from such proposals. Concern mounted and in 1979 an amendment to the Fisheries Act allowed for the creation of aquatic reserves rather than just the restriction of fishing by closures. As the area supported a thriving tourist industry and extensive fisheries, to be successful the reserve needed to cater for these existing uses within a framework of biodiversity conservation.

As with the Great Barrier Reef Marine Park, involvement of the local community was essential. A user-friendly consultation document was produced, encouraging local participation. This was accompanied by a draft zoning scheme worked out by NSW Fisheries together with sectors of the local community. Four different zones were identified in the proposals:

1. General use zone.
2. Recreation zone.
3. Refuge zone.
4. Sanctuary zone.

NSW Fisheries used community service announcements to seek the views of an estimated 120 000 people. The result of this public participation programme was a revised zoning scheme, discussed and agreed to by the local community and supported by a management plan and fisheries regulations. In May 1991 the reserve was finally gazetted under the 1979 Aquatic Reserve Provisions of the Fisheries and Oyster Farms Act.

The zoning scheme for the Solitary Islands is similar to that of the Great Barrier Reef Marine Park but far simpler, a reflection of a smaller area to manage and a smaller range of activities to be taken into account. The majority of the reserve is open to general use, with areas of restriction being small and specific to the various islands, estuaries or headlands (Plate 3). A compromise was reached in some situations, and particularly around the Solitary Islands themselves, where the consultation process identified a demand for continuing use of areas for fishing. One side of the island has been declared a refuge zone, with limited line fishing, and the remainder has been established as a non-extractive sanctuary.

The community use the zoning scheme rather than the management plan or regulations to understand how to use the area. The NSW Fisheries hopes that, armed with an understanding of the Reserve's objectives, self-policing will develop.

6.7.3 Lord Howe Island proposed Marine Reserve

Lord Howe is approximately 630 km off the north coast of New South Wales. It is a volcanic island consisting of a narrow strip of land 11 km long orientated in a north–south direction. The island is surrounded by several small islets. The Admiralty Islets group lies to the north-east and Ball's Pyramid, a spire rising 549 m from the sea, is 23 km south-west. Lord Howe Island group possesses the world's most southerly coral barrier reef and associated lagoon. In December 1982, Lord Howe Island became the first site in New South Wales to be included on UNESCO's World Heritage List.

There is currently a proposal from the New South Wales Fisheries to establish a marine reserve in the State waters around the islands which extend for 3 nautical miles from Lord Howe Island, Ball's Pyramid and the adjacent islets.

In developing the reserve proposal, NSW Fisheries has adopted a multi-zoning approach taking account of users' views. This zoning scheme allows traditional uses of the marine environment and, at the same time, protects representative or sensitive components of the marine ecosystem. Within the zoning plans, three multiple-use zones are identified:

1. General use zone.
2. Refuge zone.
3. Sanctuary zone.

This approach describes what can be done in different areas of the marine reserve and states: '. . . you will find that there are a great variety of things that you may do. The zoning plan provides for the complete protection of certain areas and the partial protection of others. This zoning scheme does not mean that you can no longer fish, although you may have to avoid fishing within some areas.'

The reserve proposal has been through the process of public consultation and it will be interesting to monitor developments once it has been declared.

6.7.4 Lundy Island Marine Nature Reserve

Lundy Island lies at the entrance to the Bristol Channel in south-west England. It is a rugged island rising nearly 140 m out of the water with spectacular scenery and wildlife. Lying north–south, it is nearly 5 km long but only 1.3 km across at its widest point.

The idea of having a marine reserve around the island originated in 1969 when divers first recorded the presence of warm-water marine species. In 1971 a proposal was made by the Lundy Field Society to establish a marine reserve. Following the Wildlife and Countryside Act 1981 (which introduced marine reserve provisions into UK legislation) and public consultation, the reserve was designated on 21 November 1986. Lundy remains England's only statutory marine nature reserve.

Over the last 3 years the management and interpretation of the marine reserve have been reviewed and new systems introduced to increase the effectiveness of the reserve. This has included the development of a zoning scheme showing the existing management arrangements (English Nature, 1993). This was developed because very few visitors or users of the reserve were aware of the management arrangements, which were only described in a draft management plan. The zoning scheme has now been adopted by the key organizations involved with managing the marine reserve, namely English Nature (the UK government's statutory advisor on nature conservation in England), the Landmark Trust (who lease the island) and the Devon Sea Fisheries Committee.

The zoning scheme has been developed using principles derived from zoning schemes in Australia to give a visual interpretation of the controls on activities in the reserve. Five zones are recognized, showing the range of use and levels of habitat protection in the reserve (Plate 4). These are:

1. General use zone
2. Recreational zone
3. Refuge zone
4. Sanctuary zone
5. Archaeological protection zone.

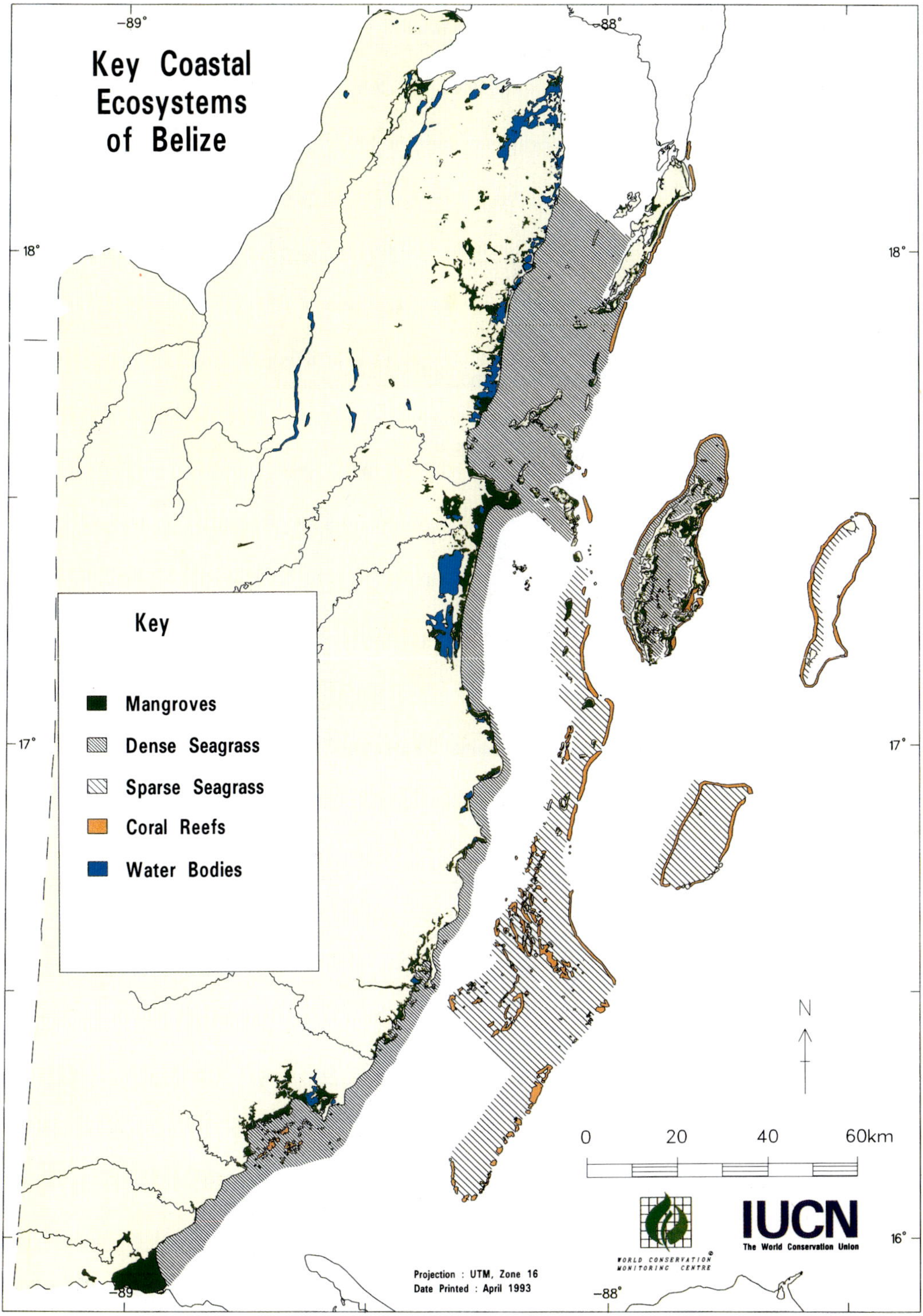

Figure 2.1 Key coastal ecosystems of Belize.

FAR NORTHERN SECTION ACTIVITIES GUIDE

ACTIVITIES	General Use 'A' Zone	General Use 'B' Zone	Estuarine Conservation Zone	Marine National Park 'A' Zone	Marine National Park Buffer Zone	Marine National Park 'B' Zone	Scientific Research Zone	Preservation Zone
Boating, diving	Yes	Yes	Yes	Yes	Yes	Yes	No	No
Collecting (e.g. beach-de-mer, shells, coral, aquarium fish	Permit	Permit	Permit	No	No	No	No	No
Line fishing	Yes	Yes	Yes	Yes	No	No	No	No
Gill netting	Yes	Yes	Yes	No	No	No	No	No
Bait netting	Yes	Yes	Yes	Yes	No	No	No	No
Trolling (for pelagic species)	Yes	Yes	Yes	Yes	Yes	No	No	No
Spearfishing	Yes	Yes	Yes	No	No	No	No	No
Pole and line tuna fishing	Permit	Permit	N/A	No	No	No	No	No
Trawling	Yes	No	No	No	No	No	No	No
Traditional fishing	Yes	Yes	Permit	Permit	Permit	Permit	Permit	No
Traditional Hunting	Permit	Permit	Permit	Permit	Permit	Permit	Permit	No
Cruise ships	Yes	Permit	Permit	Permit	Permit	Permit	No	No
General shipping (other than shipping area	Yes	No	No	No	No	No	No	No
Crayfishing	Yes	Yes	Yes	No	No	No	No	No
Scientific research	Permit	Permit	Permit	Permit	Permit	Permit	Permit	Permit

CAIRNS SECTION ACTIVITIES GUIDE

ACTIVITIES	General Use Zone	Habitat Protection Zone	Estuarine Conservation Zone	Conservation Park Zone	Buffer Zone	National Park Zone	Preservation Zone
Diving, boating, anchoring, photography	Yes	Yes	Yes	Yes	Yes	Yes	No entry
Line fishing	Yes	Yes	Yes	Limited 1	No	No	
Trolling	Yes	Yes	Yes	Pelagic species only	Pelagic species only	No	
Spearfishing (snorkel only)	Yes	Yes	Yes	Restricted sites 2	No	No	
Collecting 3	Permit	Permit	No	No	No	No	
Limited collecting 4	Yes	Yes	Yes	No	No	No	
Bait gathering (by hand)	Yes	Yes	Yes	Yes	No	No	
Bait netting	Yes	Yes	Yes	Yes	Pelagic species only	No	
Crabbing	Yes	Yes	Yes	Limit on pots 5	No	No	
Oyster gathering	Yes	Yes	Yes	Limited 6	No	No	
Commercial netting	Yes	Yes	Yes	No	No	No	
Trawling	Yes	No	No	No	No	No	
Aircraft	Yes	Yes	Yes	Yes	Yes	Yes	Above 500'

Permits are required for most other activities. This includes research, educational programs, traditional hunting, waste discharge, placement of moorings or other structures, anchoring or mooring for an extended period, tourist operations and most commercial activities.

See other side of map for detail. For clarification of any points or further information contact the Great Barrier Reef Marine Park Authority, Department of Environment and Heritage or Queensland Boating and Fisheries Patrol.

1 Maximum of 2 lines/rods per person, and 1 hook/lure per line.

2 Spearfishing may be undertaken at the following locations in the Conservation Park Zone:
-Hope Island area
-Snapper Island area
-Frankland Islands (South) area
-North Barnard Islands area.

3 Taking of more than 5 of any unprotected species in a 28 day period (excludes coral).

4 Generally taking of not more than 5 of any species of fish, shell or invertebrates.

5 No more than 4 catch devices per person.

6 Taking oysters for immediate consumption on site.

Plate 1 Zoning Box and footnotes for a section of the Great Barrier Reef Marine Park.

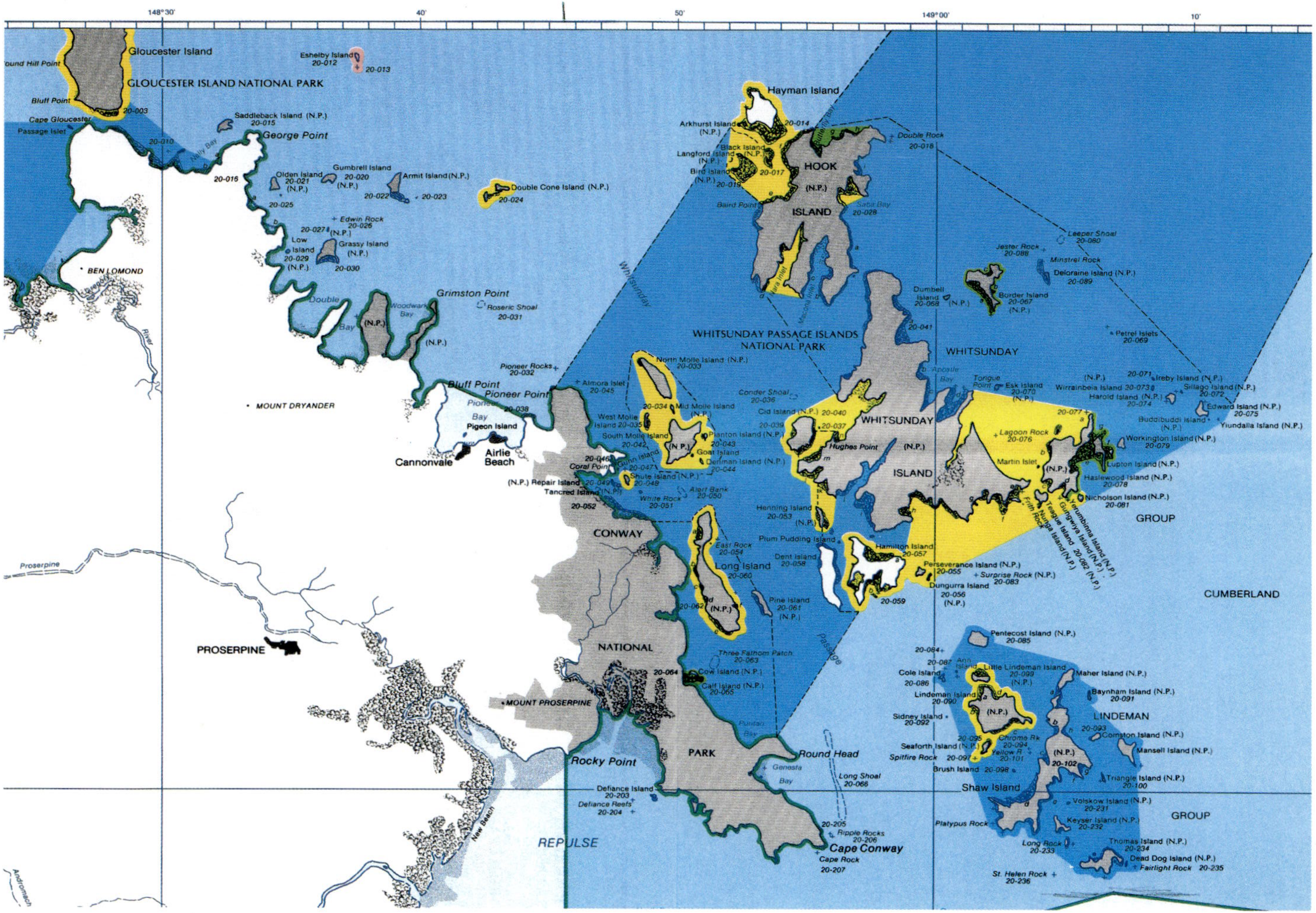

Plate 2 Whitsunday Group, one of the zoned areas of the Great Barrier Reef Marine Park.

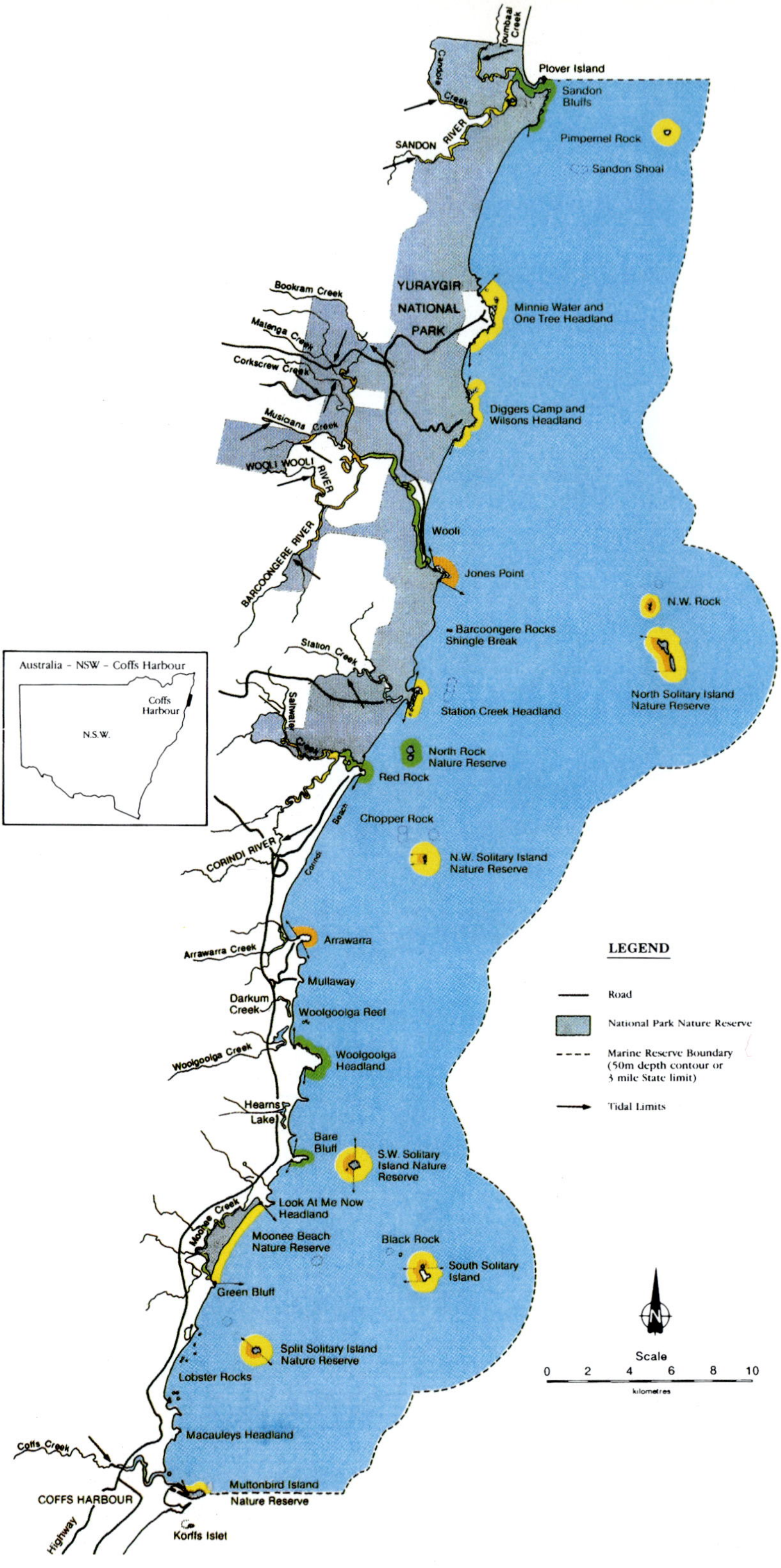

Plate 3 Zoning scheme for the Solitary Islands Marine Reserve.

SOLITARY ISLANDS MARINE RESERVE

ZONING SCHEME

ACTIVITY	GENERAL USE ZONE	RECREATION ZONE	REFUGE ZONE	SANCTUARY ZONE
RECREATIONAL				
- Line fishing	Yes	Yes	Yes	No
- Netting	Yes	Yes	No	No
- Trapping	Yes	Limited (2)	Limited (3)	No
- Spearfishing	Yes	Yes	Limited (4)	No
- Collecting	Yes	Limited (1-2)	No	No
- Diving	Yes	Yes	Yes	Limited (A)
COMMERCIAL				
- Line fishing	Licenced	Licenced	Licenced	No
- Trawling	Licenced	No	No	No
- Trapping	Licenced	Limited (2)	Limited (3)	No
- Netting	Licenced	No	No	No
- Collecting	Limited (1-2)	Limited (1-2)	No	No
- Tourist Activity	Permit	Permit	Permit	Permit
COMPETITIONS				
- Line and spearfishing	Club Permit	Club Permit	Club Permit	No
COLLECTING FOR				
- Aquariums	Permit	Permit	No	No
- Group educational excursions	Permit	Permit	Permit	No
- Scientific Research	Permit	Permit	Permit	Permit

A) No precertified diver training in Sanctuary Zones.
1) For pipis, beach worms, yabbies, cunjevoi, oysters, green sea lettuce, blackfish weed, crabs, sea urchins, prawns and dead objects only.
2) Rock lobsters, crabs and abalone only.
3) Rock lobsters in headland Refuge Zones only.
4) Spanish mackerel, mulloway, cobia, wahoo, rainbow runner, yellowtail kingfish, samson fish, amberjack, chinaman leather jacket, sawtail surgeon, bream, tarwhine, tailor, diamond trevally, golden trevally, luderick, dusky flathead, red morwong, sweetlip.

NOTE:

(i) The use of anchors in Sanctuary zones is prohibited to protect fragile corals.
(ii) These activities must comply with existing Fisheries regulations, e.g. restrictions on collecting and fishing at North Solitary Island; restrictions on trapping at North and South Solitary Islands; restrictions on spearfishing, netting and trapping in estuaries and bag and size limits applying to recreational fishers.
(iii) All zones cover intertidal areas and most extend to 200m or 400m from the low tide line. A booklet detailing all zones of the Reserve is available from:-

Solitary Islands Marine Reserve office,
P.O. Box J154
Coffs Harbour NSW 2450
Telephone: (066) 523 977

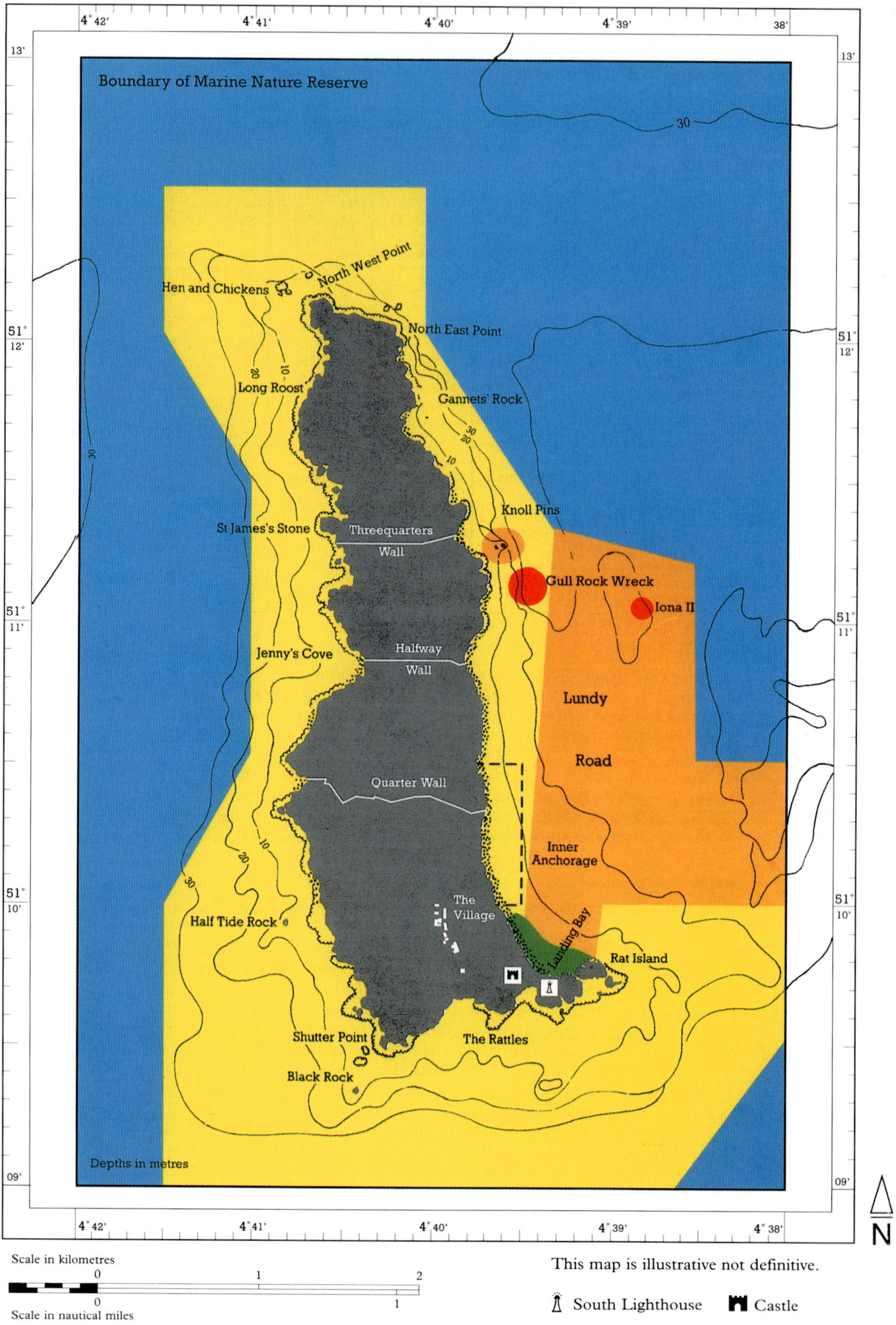

Plate 4 Zoning scheme for the Lundy Island Marine Nature Reserve.

	Activity	General Use Zone	Recreational Zone	Refuge Zone	Sanctuary Zone	Archaeology Protection Zone
Recreational	Diving	Yes	Yes	Yes	Yes	No
	Snorkelling[1]	No	Yes	No	No	No
	Swimming[1]	No	Yes	No	No	No
	Spearfishing	No	No	No	No	No
Commercial	Trawling	Yes	No	No	No	No
	Dredging	Yes	No	No	No	No
	Potting	Yes	Yes	Yes[2]	Limited[3]	No
	Tangle nets	Yes	No	Limited[4]	No	No
	Fixed nets	Yes	No	Limited[4]	No	No
Collecting	Group educational excursions	Permit	Permit	Permit	Permit	No
	Scientific research	Permit	Permit	Permit	Permit	Permit

1 The Landing Bay has a snorkel trail, where it is safe to swim or snorkel at low water. There are many hazardous currents and steep cliffs around the island, so seek advice on the island before snorkelling and swimming elsewhere. All these types of activity are undertaken at your own risk.

2 Potting is not encouraged in a 200m zone stretching seaward from high water mark, from Gannets' Rock South to the Landing Bay.

3 You will require a permit for potting between 1 October and 30 June in the Sanctuary Zone. Potting is prohibited within 100m of the low water mark around the Knoll Pins.

4 You can use tangle and fixed nets in only one part of the Refuge Zone and then only if you have a licence. This is the area west of the line running due north on longitude 04°39.5'W between the points 51°10'N and 51°10.5'N (area within dashed line on map), as far as the mean high water mark medium spring tides.

Please Note -

(i) All vessels are asked to avoid anchoring in all zones whenever possible, except the general use zone and near the Landing Bay in the Inner Anchorage. Anchoring within the archeological exclusion zone is not allowed at any time.

(ii) All activities have to comply with existing fisheries regulations and byelaws (Devon Sea Fisheries Committee byelaws 13 to 15 give further information). You are discouraged from rod and line fishing in the Landing Bay area because of the danger to snorkellers and swimmers from fishing tackle. The wardens can advise on suitable locations.

(iii) Where zones touch the shore, they cover all intertidal areas within the reserve up to the point of the highest astronomical tide.

The Warden for the Marine Nature Reserve can be contacted for advice and permits on (01237) 431831

Apart from summarizing the management arrangements on a chart, the zoning scheme serves another important function. In the UK there has been continued debate and concern during the last few years about the lack of integration concerning management of activities in coastal waters. There is no overarching coastal zone management authority and management responsibilities are divided between many different sectoral interests. The Lundy zoning scheme provides a user-related framework which gives an overview of the actions of different sectors and, for the first time in the UK, has provided a common framework to express, in this particular case, environmental protection afforded by fisheries, archaeological and conservation legislation.

A criticism of the zoning scheme is its complexity for such a small area but this only highlights the need to present management information in an accessible and easily understandable format. When this is done it can reveal whether the proposed arrangements are too complex. Revision of the zoning scheme will start in 1996/7 when the aim will be to simplify the overall structure and establish more common-sense boundaries to the various zones.

6.8 CONCLUSIONS

The four case studies highlight the central role that zoning schemes can play in the development and implementation of marine protected areas. The framework that such schemes provide helps to resolve many of the complexities of management in areas of overlapping jurisdiction in the marine environment. Together with appropriate interpretation and publicity material, they are also an unparalleled link to the users of such areas. Providing such a high potential for understanding in users helps in the difficult task of encouraging self-policing of MPAs.

The next steps, at an international level, are for a wider evaluation of existing zoning schemes and the development of accepted standards for individual schemes. A consistent message, in terms of both development and deployment, and in the use of colour interpretation of uses and legislation, should heighten further the value and application of this aid to the management of marine protected areas.

REFERENCES

Batisse, M. (1990) Development and implementation of the biosphere reserve concept and its applicability to coastal regions. *Coastal Conservation*, **17**(2), 111–116.

English Nature (1993) *Managing Lundy's marine wildlife. Have your say!*, English Nature, Peterborough, 6 pp.

Hopley, D. (1989) *The Great Barrier Reef: ecology and management*, Longman Cheshire Pty Limited, Sydney, 53 pp.

Kelleher, G. (1993) *A model for the management of coastal areas – the Great Barrier Reef Marine Park*. Opening lecture to the Baltic Sea Region Seminar on Establishment, Protection and Effective Management of Coastal and Marine Protected Areas in the Baltic Sea Region, Nykoping, Sweden, 8–10 June 1993.

Kelleher, G. and Kenchington, R. (1992) *Guidelines for establishing marine protected areas*, IUCN, Gland, Switzerland, 79 pp.

Laffoley, D. d'A. (1994) *The establishment, development and promotion of marine reserves in Australia and New Zealand. A report to the Winston Churchill Memorial Trust*, English Nature, Peterborough, 20 pp.

Laffoley, D. d'A, Baxter, J.M., Connor, D.W., Gilliland, P., Hiscock, K., Platt, H., Richards, M. and Tasker, M. (1994) The development of a generic management model for European Union marine Natura 2000 sites. In *Marine Protected Areas and Biosphere Reserves: Towards a new paradigm* (ed D. J. Brunckhorst). Proceedings of the 1st International Workshop on Marine and Coastal Protected Areas, Canberra, Australia, August 1994, 98 pp.

Price, A. and Humphrey, S. (eds) (1993) *Application of the Biosphere Reserve Concept to Coastal Marine Areas*. Papers presented at the UNESCO/IUCN San Francisco Workshop of 14–10 August, 1989. A Marine Conservation and Development Report, IUCN, Gland, Switzerland, viii + 114 pp.

Salm, R.V. and Clark, J.R. (1984) *Marine and coastal protected areas: a guide for planners and managers*, IUCN, Gland, Switzerland.

Sherman, K. and Laughlin, T. (1992) *The Large Marine Ecosystem (LME) Concept and its application to Regional Marine Resource Management*. Marine Conservation and Development Reports, IUCN, Gland, Switzerland, vi + 62 pp.

UNESCO (1987) *A Practical Guide to MAB*.

7

Enforcement in marine protected areas

Billy D. Causey

Enforcement is an essential component in the management of marine protected areas (MPAs) but the level of enforcement required can vary significantly from site to site and from programme to programme. Enforcement policies, authorities and philosophies differ between agencies and programmes within nations (e.g. the National Park Service and National Marine Sanctuaries programme in the United States) and, to an even greater degree, between international programmes. Traditionally, enforcement has relied on citations and penalties as forms of punishment to establish compliance and discourage violators. This is particularly true of crimes against the State or society in municipalities around the world.

Although this form of enforcement is necessary under some circumstances, education can lead to greater overall compliance in a shorter timeframe in MPAs, and its effects can be longer lasting. Salm and Clark (1984) suggested that the most effective means of ensuring the compliance of legislation is through public education and enlisting the help of user groups in management. Hudson (1988) believes that the best forms of enforcement are education and peer pressure. These views are similar to the philosophy of 'interpretive law enforcement' used by the United States National Marine Sanctuary Program in the Florida Keys (Causey, 1989). This form of enforcement is very effective as it can generate both public support and visitor compliance. It is therefore critical to recognize that education and interpretation are essential tools in any enforcement effort for MPAs. Chapter 9 looks at the role of education in MPAs in some detail. In this chapter the case studies illustrate how education and other measures have been used to support the enforcement of regulations in two MPAs in the United States.

7.1 NEED FOR ENFORCEMENT

Effective management of MPAs relies on resource protection, restoration, wise use, education, interpretation and enhancement. Designating an MPA without providing some form of management to protect the resources can

Marine Protected Areas: principles and techniques for management. Edited by Susan Gubbay.
Published in 1995 by Chapman & Hall, London. ISBN 0 412 59450 1.

have less than beneficial results. Despite this, legislation is enacted and MPAs are designated to protect an area all too frequently without a mechanism to enforce the basic regulations that are there to provide this protection. According to an inventory of Caribbean marine protected areas (OAS, 1988), approximately 135 marine and coastal protected areas were established in the Greater Caribbean Basin by 1988. Of these, only 16% of those outside the jurisdiction of the United States were 'fully protected' under the criteria used by the Organization of American States. These statistics suggest that far more areas were being designated than were receiving the management commitment it takes to protect the resources from the primary use, which in this region is tourism, even though the motivating factor for establishing these protected areas was to safeguard the natural resources that attracted visitors.

Regardless of the reasons for designating MPAs, it is essential that a management plan and all its components – education, interpretation, research and enforcement – are developed early and set out how the objectives of the reserve are to be met. If visitor compliance and local acceptance of the rules and regulations of the MPA are not achieved, then there has to be some capacity to nudge people in the direction of compliance through 'interpretive enforcement', an approach which relies heavily on education and contact with the public. If interpretive enforcement does not accomplish the task, it may be necessary to initiate a more aggressive enforcement approach. An enforcement presence on the water, for example, may be needed from the outset to protect the natural resources of the site.

The title 'marine protected areas' implies that some element of the site's management plan provides rules and regulations designed to protect the resources. This suggests that there should be an enforcement programme, whatever its approach or philosophy, to enforce those rules and regulations. This commitment is often overlooked or ignored in the designation of MPAs and is one of the main reasons that MPAs managed on paper do not achieve their very basic management objectives. The concept of enforcement as a management tool in MPAs is often neglected or given low priority by upper levels of management. It is not sufficient to talk or write about management strategies and action plans. If some form of enforcement is not available to guarantee that visitors will comply with the regulations, then the public can lose respect for the programme, and those opposed to the designation of the MPA can develop an even greater level of contempt for 'government control'.

When a marine area is set aside for protection because of its natural, cultural, aesthetic or economic value there is an obligation and responsibility to enforce the regulations. The precise form this takes will depend on the philosophy of the enforcement programme, the degree of enforcement authority and the levels of enforcement effort. This will be site specific even within a single national programme that establishes and manages MPAs. Generally speaking, if the MPA receives lots of visitors and allows multiple uses, including consumptive or extractive activities, then the required level of enforcement will be greater than for a small, discrete MPA that focuses on protecting a cultural or historical resource that receives few visitors.

The first site designated within the United States National Marine Sanctuary Program was the wreck of the Civil War ironclad warship USS *Monitor*. This ship, which sank in 67 m (220 ft) of water 26 km (16 miles) south-south-east of Cape Hatteras Light, North Carolina, is closed to visitors except by permit. The regulations for the *Monitor* National Marine Sanctuary protect the historically important site from activities that could lead to the destruction or loss of its cultural/historical artefacts. Considering the remoteness of the site and the regulation that prohibits vessels from even stopping over the protected area (0.25 square nautical miles), the need for enforcement has been minimal.

In contrast, the coral reefs protected by the Key Largo NMS were originally set aside for protection due to pressure from tourism. Proponents for the establishment of the park identified the tourist boom of the late 1950s as being largely responsible for the depletion of the coral reef resources. The growing popularity of scuba-diving was a new form of impact to the reefs. Spear-fishing, queen conch harvesting, tropical fish collecting and the demand for souvenirs of coral and shells by the tourist trade were so intense that in 1957 a group of concerned environmentalists, led by the late Dr. Gilbert Voss, requested immediate government action to protect the reefs off Key Largo. This plea was answered with the dedication of the John Pennekamp Coral Reef State Park in December 1961. With so many activities putting pressure on the park, a much higher level of enforcement was needed.

The following sections discuss issues relating to enforcement with particular reference to the United States National Marine Sanctuary Program.

7.2 ENFORCEMENT IN UNITED STATES NATIONAL MARINE SANCTUARIES

Title III of the Marine Protection, Research, and Sanctuaries Act 1972, as amended, provides the authority for the National Marine Sanctuary Program in the United States. The programme is administered by the Sanctuaries and Reserves Division of the National Oceanic and Atmospheric Administration (NOAA), part of the US Department of Commerce. It was established to designate and manage nationally significant marine areas by regulating human activities that may degrade or jeopardize the integrity or health of the resources in those areas. Research, education and management programmes are established to increase scientific understanding of the areas and to inform the public of the need to use the marine resources wisely, thus lessening adverse impacts on sanctuaries.

The National Marine Sanctuary Program allows for multiple uses, including some consumptive or extractive activities but with the primary purpose of resource protection. This is favoured by some user groups (such as recreational and commercial fishermen) as it allows them to continue in their harvesting practices, while resource protection efforts are left to the managers of the sanctuary. The success of such management depends on a variety of factors, including enforcement of sanctuary regulations that address harmful fishing practices and the concurrent enforcement of state and/or federal fisheries regulations that are established independently of the regulations for the sanctuaries. The programme emphasizes the sustaining capability of the oceans if human use is managed and impacts are eliminated or successfully mitigated (Christhilf, 1986).

Each of the 13 national marine sanctuaries is unique but to achieve the priority goal of long-term protection of the resources early decisions have to be made about the level and type of management and enforcement back-up that will protect the resources adequately. To cover this the designation process for each site provides for the development of management plans that address enforcement of regulations as an integral part of the programme management. The regulations for each sanctuary are published in the US Code of Federal Regulations.

The National Marine Sanctuary Program does not have an enforcement body under its direct administration but there are a number of Sanctuary Agents assigned to the programme from National Marine Fisheries. Enforcement is achieved through cooperative agreements with other federal and state agencies. Between 1975 and 1980, the programme had only two designated sites: the Key Largo and *Monitor* National Marine Sanctuaries. Of these, only Key Largo required an enforcement presence during that period and enforcement was accomplished through a cooperative agreement between NOAA, the State of Florida Department of Natural Resources (Division of Recreation and Parks) and the US Coast Guard.

Four national marine sanctuaries were designated in 1980–81. Among these were the Looe Key Marine Sanctuary which together with the Key Largo NMS designated in 1975 have since been incorporated into the Florida Keys National Marine Sanctuary, which encompasses 2800 square nautical miles from Miami to the Dry Tortugas. The Congressional Act that designated Florida Keys NMS was signed into law on 16 November 1990.

As the new combined sanctuary is in the early stages of development with detailed procedures still to be worked out, the following case study looks at the enforcement procedures that used to be in place for the two smaller sanctuaries rather than those for the whole Florida Keys.

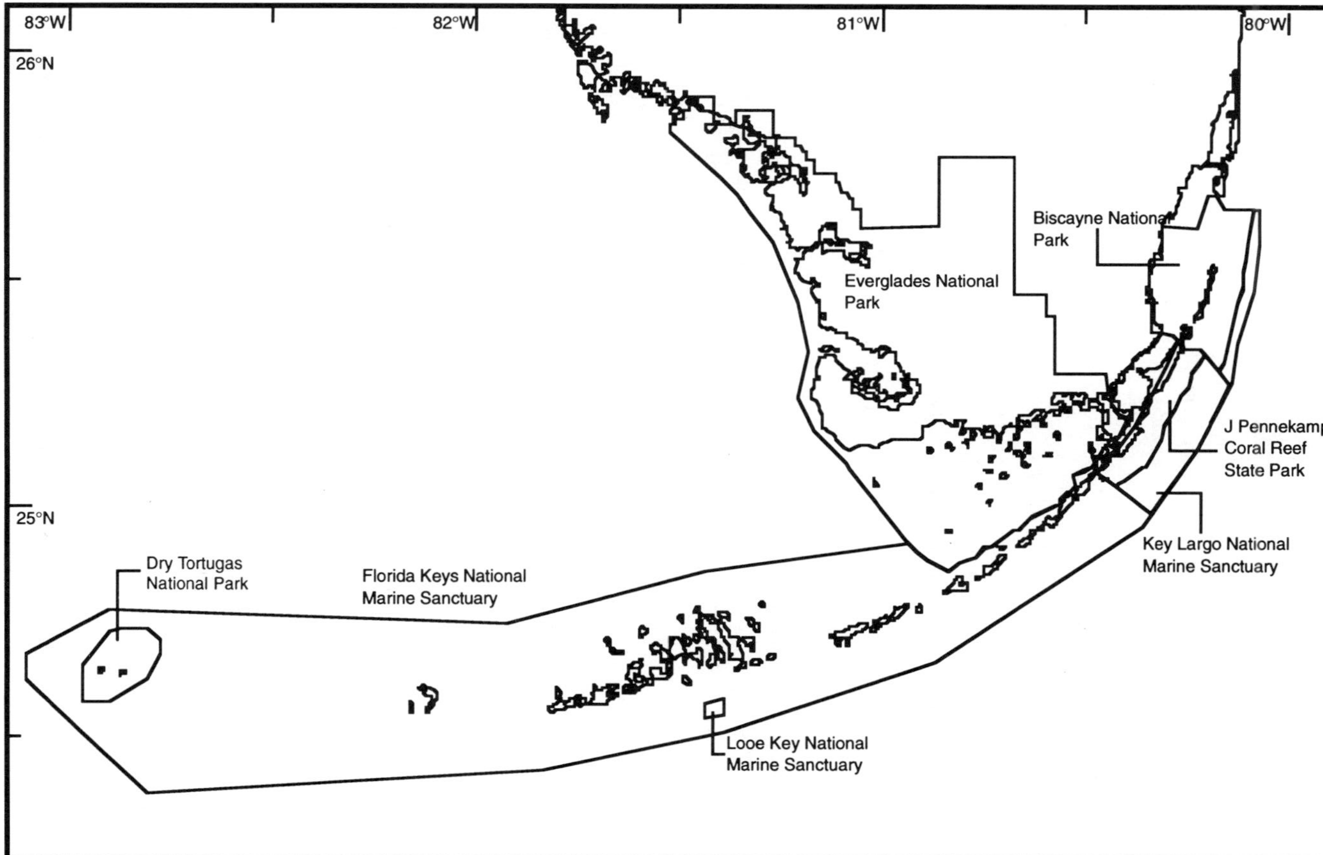

Figure 7.1 Location of the Florida Keys National Marine Sanctuary.

7.3 KEY LARGO AND LOOE KEY NATIONAL MARINE SANCTUARIES: A CASE STUDY OF ENFORCEMENT IN MARINE PROTECTED AREAS

The Florida Keys form an arc of small islands that curve south-westerly for approximately 352 km from the southern tip of the south Florida mainland. These Keys lie between the Gulf of Mexico and the Straits of Florida in the Atlantic Ocean. The combination of climate, water clarity, ocean currents, substrata and water depth has provided the conditions that allowed the formation of the Florida Keys coral reef ecosystem (Figure 7.1). This coral reef is the only one that lies immediately adjacent to the continental United States and is the third largest barrier reef in the world. Because of the Florida Keys reef tract, south Florida has become the vacation destination of hundreds of thousands of tourists annually who can drive their cars to the 'islands'.

The two most similar enforcement programmes in the National Marine Sanctuary Program were those of Key Largo and Looe Key, both in the Florida Keys. These MPAs have now been incorporated into the larger Florida Keys NMS but, because of 10 years of enforcement experience prior to this incorporation, the two sites are useful for illustrating the sorts of methods which can be used in an enforcement programme for an MPA.

As might be expected, both Key Largo and Looe Key sanctuaries protected similar resources, and were subject to similar stresses, impacts and user activities. More than a million visitors came to these sanctuaries in 1990, most of whom were divers and snorkellers who were attracted to the area because of the beauty and diversity of the coral reefs. Other visitors were fishermen who came because of the abundant marine resources. Regardless of their specific interest, visitors were attracted to the area because of what the sanctuaries had to offer in the way of natural resources. The role of management in protecting and enhancing the resources of the sanctuaries therefore became increasingly important to the economy of the area.

7.3.1 Enforcement authority

In 1982, the National Marine Sanctuary Program decided to strengthen its cooperation with the State of Florida and with the US Coast Guard. The 1982 agreement provided funding from the federal sanctuary programme to the state of Florida for various management-related activities in the two sanctuaries, including funding for all law enforcement activities at both Looe Key and Key Largo NMSs. Although the cooperative agreement was between NOAA and the Florida Department of Environmental Protection (FDEP), the responsibility for managing the agreement within FDEP was delegated to one of the department's units, the Bureau of Sanctuaries and Research Reserves (which also manages the Estuarine Research Reserve Programme for NOAA in Florida).

The cooperative management described above is an interesting state/federal partnership that benefits both partners. Although both Florida sanctuaries were in federal waters, the health and well-being of the sanctuary resources affected the economy of the state. NOAA's partner thus had a vested interest in the successful management of the resources and was able to manage the sanctuaries effectively, without creating a bureaucratic burden on the sanctuary programme. In addition, with both agencies focusing on the goals, it was a chance to leave egos and logos out of resource management.

Authority of sanctuary officers

The sanctuary officers employed by FDEP were appointed as US enforcement officers of NOAA. They were authorized to enforce all regulations established for the sanctuaries and published in the Federal Register. Each sanctuary officer was a sworn state law enforcement officer with FDEP, which requires certification by Florida's Criminal Justice Standards and Training Commission. New recruits are required to attend a 20-week law enforcement training course given by the Florida Marine Patrol (another division within FDEP).

On satisfactory completion of their training and receipt of their commission, the sanctuary officers have arrest powers that extend throughout the state of Florida. However, FDEP's commitment to the cooperative agreement (which pays 100% of the officers salaries) results in officers who are dedicated to full-time enforcement in the sanctuaries.

Unlike enforcement staff in most MPAs around the world, the sanctuary officers at Looe Key and Key Largo NMSs were able to bear firearms. This was often cause for comment, especially from foreign visitors, but was consistent with the sanctuary officers being sworn state police officers and subject to departmental guidelines. This requires law enforcement officers to be armed while performing their duties, particularly while operating vessels and vehicles that are equipped with emergency sirens and lights. Considering that south Florida has sizeable smuggling activities, carrying firearms was a useful policy for sanctuary officers and other law enforcement officers involved in water patrols in the area. This communicated a message of high-profile enforcement to visitors accustomed to the 'Park Ranger' image, which may not be desirable if the officer is encountering a family enjoying the sanctuary but may be highly desirable if the officer is confronting a poacher. To overcome this it was important for the officer to set visitors at ease when no violation had been committed. This was a fundamental enforcement philosophy of the sanctuaries and remains the case in the new Florida Keys NMS. It is discussed in greater detail below, along with the on-water patrol activities.

Prior to 1990, sanctuary officers were only authorized to enforce regulations that were specific to each sanctuary. As such, they did not have authority to enforce the National Marine Fisheries regulations established by the Magnuson Fishery Conservation and Management Act 1976, which was established for managing certain fisheries in the exclusive economic zone adjacent to the seaward boundary of the state. The sanctuary officers now have the authority to enforce National Marine Fishery regulations, as well as the Endangered Species Act of 1973 and the Marine Mammal Protection Act of 1972. The National Marine Fisheries regulations pertain primarily to fishery-related management, such as closed seasons, size limits, catch limits, gear restrictions and the protection of endangered or threatened species. This additional authority benefits sanctuary management enormously by giving the sanctuary officers the authority to enforce fishery regulations established to protect marine resources from over-harvest. Furthermore, the officers are capable of enforcing these fishery regulations both inside and outside the boundaries of the sanctuaries.

Level of enforcement

Three sanctuary officers were assigned to the Looe Key NMS and four to the Key Largo NMS. Looe Key was much smaller (18.2 km^2 compared with Key Largo at 259.2 km^2) and, as a result, received more enforcement coverage. The scheduling of water patrols was crucial to maximizing surveillance during high-use periods (Figure 7.2) and, to assist with this, information was collected on visitor use.

The census form required sanctuary officers, during the course of their patrols, to identify all types of visitor use in a specific period. At Looe Key this was completed on a daily basis from 1983 (Figure 7.3) (Causey, 1989). There was also space on the form to allow officers to fill in weather data, duration of census, sea conditions, water visibility, recreational and commercial activities, research activity, educational groups and their daily activity (e.g. interpretative contacts, violations observed, enforcement actions). This helped to determine visitor-use patterns, including total number of visitors, seasonal peaks of high use, seasonal declines, types of visitor activities, changes in holiday visitation and the influence of weather. Similar data were collected at Key Largo NMS but, due to the larger area of the sanctuary compared with Looe Key NMS and the small number of sanctuary officers, the information was of limited value except for determining general visitor-use trends and patterns.

Figure 7.2 Sanctuary officers conduct surveillance activities in the sanctuary on a regular basis.

Figure 7.3 As part of their routine duties, sanctuary officers complete a daily visitor-census form.

Once visitor-use trends were determined, water patrols could be scheduled to allow for maximum enforcement surveillance during peak periods. The trends were also used to cancel patrols when conditions were likely to result in low visitation, allowing officers time for other duties. Another benefit was that the information helped in scheduling staff vacations, training sessions and regular days off work.

Additional enforcement manpower was available from other agencies in the event of a major catastrophe such as a ship grounding, or during routine operations such as pulling abandoned lobster traps after the close of the season. Several Florida Park Service Rangers assigned to John Pennekamp Coral Reef State Park and Bahia Honda State Recreational Area were deputized to enforce sanctuary regulations at Key Largo and Looe Key NMSs. These personnel could assist with enforcement calls when a sanctuary officer was not available. This arrangement worked both ways, with sanctuary officers available to assist in the Parks. Other agencies that provided support were Florida Marine Patrol, US Coast Guard, National Marine Fisheries, National Park Service and Monroe County Sheriff's Office.

The sanctuary managers responsible for supervising the law enforcement staff also needed some training in enforcement. They needed, for example, to have enough enforcement experience in order to verify that cases were properly prepared and documented by the enforcement staff and that proper procedures had been followed. However, it was important that the role of the manager was not confused with the role of the uniformed sanctuary officers in the eyes of the public. The value of this was demonstrated on several occasions at Looe Key NMS when the manager received tips from various concerned citizens that were not communicated directly to enforcement staff. This adds another dimension to the enforcement as some people communicate more freely with non-uniformed management staff rather than enforcement officers.

7.3.2 Sanctuary regulations

The regulations for each sanctuary were based on an evaluation of the resources, activity levels, the adequacy of the long-term protection provided by the existing regulatory system and the economic impacts of new regulations. Authority for promulgating sanctuary regulations is provided by Article 303 of the Marine Protection, Research, and Sanctuaries Act of 1972 (as amended) which provides that the Secretary of Commerce, after consultation with other interested federal agencies, shall issue necessary and reasonable regulations to control activities permitted within the designated marine sanctuary (Sanctuaries Program Office, 1982).

In general, the regulations established for the Florida sanctuaries protected the coral reefs from various types of physical impact and eliminated several types of harmful or depleting fishing techniques. Since the two Florida sanctuaries accommodated high levels of recreational use, some of the regulations were designed to protect the visitors from activities that may be harmful. Although most of the prohibited activities were similar for both Key Largo and Looe Key, the establishment of a Core Area around the shallow Fore Reef habitat, to protect it from lobster trapping and boat anchors, was added to the Looe Key regulations. The Core Area at Looe Key NMS was approximately 0.5 km^2 and represented the National Marine Sanctuary Program's first attempt at zoning.

A complete list of the regulations is included in Appendices 7.A and 7.B; Table 7.1 gives abbreviated listings for each sanctuary. New regulations are being formulated for the broader Florida Keys NMS and the concept of marine zoning will be considered for use as a primary management tool. Looe Key and Key Largo NMSs will become Existing Management Zones in the Florida Keys NMS.

7.3.3 Prosecution of violations

Infractions of sanctuary regulations are civil violations and are prosecuted as civil actions. There are several advantages over prosecuting as criminal cases. The amount of evidence does not have to be as great in civil cases, for example, and it is not critical for enforcement officers actually to witness the infraction. Both of these are particularly important in the enforcement of regulations in MPAs where much of the activity is underwater and often out of sight of the sanctuary officers.

Sanctuary officers in Key Largo and Looe Key documented a violation with an **Enforcement Action Report** (EAR) (Figure 7.4). This was not a formal statement of charges, but a brief synopsis of information that serves to document the case or incident. It notified the violator that a civil action may be forthcoming. At the time of the incident, the sanctuary officer had the discretion of making the case a written warning or a sanctuary violation (which carried civil penalties). Situations in which written warnings were issued initially, nonetheless, could be prosecuted as violations in some cases. Circumstances that warranted such action included infractions committed by a violator with a history of violations, or if evidence (such as resource damage) was not observed during the initial investigation.

Offence Investigation Reports (OIRs) were used to document more serious violations, such as boat grounding. They allowed for a narrative description of the events and circumstances surrounding the infraction. The OIR and a copy of the EAR were submitted to the NOAA General Counsels Office for prosecution. If the violation involved resource damage that needed to be assessed by the sanctuary biologist (such as a boat grounding), this assessment was submitted with the OIR and EAR. The NOAA staff

Table 7.1 Abbreviated listing of prohibited activities in Key Largo and Looe Key NMSs (full details are given in Appendices 7.A and 7.B)

Key Largo	*Looe Key*
1. Removal or damage of natural features and marine life (except the taking of spiny lobster and stone crab allowed under the appropriate Fishery Management Plan). The collection of tropical fish and invertebrates is prohibited under this regulation.	1. Removal or damaging of distinctive natural features and marine life (except the taking of spiny lobster under the appropriate Fishery Management Plan). This includes the collection of tropical and marine invertebrates.
2. Dredging, filling, excavating and building activities.	2. Anchors or anchoring apparatus touching coral in the Core Area.
3. Discharge or deposit of refuse and other pollutants.	3. Failure to use available mooring buoys.
4. Tampering with markers, mooring buoys and scientific equipment.	4. Discharge of refuse and other pollutants.
5. Use of harmful fishing methods including: spear-guns; spearing devices; wire fish traps; bottom trawls; dredges; anchored bottom fishing gear or nets; poisons; electric charges; explosives or similar devices.	5. Tampering with markers, mooring buoys and scientific equipment.
6. Damage caused by watercraft and boat anchors.	6. Use of spear-guns and other spearing devices.
7. Handling and standing on coral formations.	7. Use of wire fish traps and other harmful fishing techniques such as poisons, electric charges and explosives.
8. Removal or destruction of archaeological, historical or cultural resources.	8. Operation of watercraft in such a manner as to strike or otherwise cause damage to corals and other natural features of the sanctuary.
9. Operation of watercraft in excess of 4 knots within 100 yards of a 'divers down' flag.	9. Handling and/or standing on coral formations.
10. Failure to fly a 'divers down' flag.	10. Setting of lobster traps within the Core Area.
11. Use of dangerous weapons such as explosives or firearms.	11. Failure to fly a 'divers down' flag.
12. Failure to use available mooring buoys.	12. Removal or destruction of archaeological, historical or cultural resources.

attorney then took these reports and issued a **Notice of Violation Assessment** (NOVA) to the party or parties deemed responsible for the violation. The responsibility for damages could extend to parties that were not present at the scene of the incident when it occurred. This is common in the case of

Page 1 of Pages

NOAA FORM 88–151 (6–78) U.S. DEPT. OF COMM. NOAA – NMFS

ENFORCEMENT ACTION REPORT

CCN: *(Regional Office use only)*

NAME OF VIOLATOR (Last - First - Middle) | DATE OF BIRTH

ADDRESS OF VIOLATOR

NAME OF VESSEL | NATIONALITY

INTL. RADIO CALL SIGN/DOCUMENTATION NO. | DATE OF VIO.

LATITUDE/LONGITUDE/PLACE OF VIOLATION

SEIZED PROPERTY RECEIPTS
YES ☐ # ______________ NO ☐

COUNT	TYPE OF ACTION	REGULATION/STATUTE VIOLATED

AUTHORIZED OFFICER | OFFICER'S ID NO.

DUTY STATION/VESSEL

DATE AND TIME OF REPORT | E.A.R. NO. **A** 40876

CASE FILE COPY

OPTIONAL INFORMATION

PHYSICAL DESCRIPTION OF VIOLATOR

HEIGHT	WEIGHT	COLOR OF HAIR	COLOR OF EYES
SEX	RACE	CITIZENSHIP	

ADDITIONAL VESSEL INFORMATION

TYPE	LICENSE NO.
L.O.A.	G.T.
COLOR	HOME PORT/HAILING PORT

REGISTERED OWNER'S NAME AND ADDRESS

ADDITIONAL INFORMATION OR REMARKS

Figure 7.4 Enforcement Action Report.

grounding of vessels rented or leased to the violator by a second party, and where the boat is owned by a third party. All parties could be jointly or severally charged for damage to the resources.

The OIR was also used to document violations not witnessed by the sanctuary officers. In civil cases, such as those recorded in the sanctuaries, the information provided by a witness to the violation could be used to prosecute the case.

7.3.4 Penalties for violations

A penalty schedule was established by the General Counsel's Office for each of the Florida sanctuaries and will be developed for the Florida Keys NMS. It was used to establish penalty amounts if a NOVA was issued and to establish summary settlement amounts for minor violations (Table 7.2). Table 7.3 lists various sanctuary regulations and an approximate penalty range for each violation for first, second and third time violators. The Marine Protection, Research, and Sanctuaries Act of 1971 specifies that a civil penalty up to $100 000 per day per violation can be assessed.

A sanctuary officer could make the determination that a violation was not serious. These were usually routine violations such as spear-fishing, anchoring on coral, or some slight physical contact with the corals. In these cases the officers could use the summary settlement schedule and record the settlement amount (fine) on the citation. The violator was given 20 days to mail the fine to the Office of the General Counsel. Sanctuary officers did not per-

Table 7.2 Summary settlement schedule, Looe Key Sanctuary Violations under 15.CFR Part 937

Description of violation		*Summary settlement amount ($)*
1. Handling or standing on coral		25.00
2. Non-commercial taking of marine molluscs (shells)	1–2 shells:	50.00
	3–4 shells:	75.00
	5–6 shells:	100.00
3. Non-commercial taking of marine invertebrate (e.g. sand dollars, sea biscuits, shrimp)	1–5 items:	50.00
	6–10 items:	100.00
4. Non-commercial taking of tropical fish	1–10 tropicals:	50.00
	11–20 tropicals:	100.00
5. Minor, non-aggravated anchoring on, or causing damage to, coral in Fore Reef		150.00
6. Minor vessel grounding		150.00
7. Failing to fly the 'divers down' flag		25.00
8. Using a spear-gun or sling	0 food fish harvested:	50.00
	1–2 food fish harvested:	100.00
	3–4 food fish harvested:	150.00
	5–6 food fish harvested:	200.00

sonally collect fine monies. The cases that could be settled in this manner were very specific and could be easily assessed or quantified by the officer. The option of a summary settlement schedule saves a considerable amount of paperwork and helps expedite cases.

7.3.5 Vessel groundings

This section on vessel grounding refers only to assessments relative to administrative penalties as opposed to natural resource damage action filed in US District Courts which (along with civil forfeiture of offending vessels) are authorized by the Marine Protection, Research and Sanctuaries Act of 1972.

Small boat and ship groundings on the reefs of the Florida Keys have been recognized as a major impact on the coral reef resources (Jaap, 1984; Tilmant, 1987; Miller, 1988; Causey, 1990). More than 300 recorded vessel groundings took place in the Key Largo NMS between 1980 and 1992 and another 94 in Looe Key NMS between 1981 and 1992. Sanctuary officers investigated all reported and witnessed vessel groundings in the sanctuaries and enforced the regulations that prohibited damage to the natural features of the sanctuaries from boating accidents. Citations could be written for boats running aground in all habitats, including sea-grass beds, fossilized rubble bottom, hard bottom communities and coral reef communities. In the early days of the active enforcement programme at the two sites (1982–83) it was common to encounter a captain who had run aground and who was surprised to receive a citation for damages to the coral reef resources. This was probably because, prior to sanctuary enforcement, attention to boat groundings had focused on the unfortunate mishap for the captain and the resultant damage to his vessel, rather than on resource damage. This situation is frequently encountered when enforcing marine laws both inside and outside MPAs. In the case of accidents on land in the United States, there is seldom any doubt that the person at fault is responsible for damages to his property and the resources or property that they destroyed. They can also be fined for their negligence. Should damage to natural resources resulting from negligence be treated any differently?

More recently when sanctuary officers responded to groundings, visitor safety remained a top priority, but attention was focused quickly on the damage to the resources and ways of getting the vessel free of the reef without causing any additional or unnecessary damage to the resources (Causey, 1990).

Litigation of grounding cases consumed a considerable amount of staff time over the years. To expedite grounding cases and establish a consistent method by which civil penalties could be assessed, a procedure was set up by the General Counsel's Office. This included a biological assessment of

Table 7.3 Approximate penalty range for Sanctuary violations in the Key Largo and Looe Key National Marine Sanctuaries

Type of violation	*First offence*	*Second*	*Third +*
Damaging or taking any coral or marine invertebrate	$100–2500 or damage assessment	$2500–5000 or damage assessment	$5000 + or damage assessment
Handling of coral or standing on coral formations by divers	$25–150	$150–1000	$500 +
Taking tropical fish or marine invertebrate, except as incidental to allowed fishing activities	$100–2000	$1000–5000	$2000 +
Anchoring on coral, or injuring coral with any chain or rope	$300–1000 or damage assessment	$600–2000 or damage assessment	$1200 + or damage assessment
Failure of first diver to properly secure anchor; or other diving before anchor is set	$25–150	$100–500	$500–2000
Failure to use available mooring buoy	$50–100	$100–200	$200 +
Operating vessel so as to strike or damage natural features			
– boat grounding	$300–50 000 (per day and per violation) (Note: This sum applies only to civil penalties, and not resource damage actions filed in US Courts.)		
– other types of minor vessel-inflicted damage	$200–1000 or damage assessment	$1000–5000 or damage assessment	$5000 + or damage assessment
Failure to fly the 'divers down' flag	$25–500	$50–1000	$250 +
Using or placing wire fish traps within the Sanctuary	$1500–2500 plus $250–500 per trap	$2500–3500 plus $500–1000 per trap	$3500–5000 plus $1000 + per trap
Placing lobster traps within the Fore Reef area (Looe Key)	$500–1000 plus $1000–1500 per trap	$1000–2500 plus $1500–3000 per trap	$2500–5000 plus $3000 + per trap
Spear-fishing within the Sanctuary			
0 fish harvested	$100–300	$300–800	$800 +
1 or 2 fish (harv)	$300–600	$600–1000	$1000 +
3 or 4 fish (harv)	$400–800	$800–1000	$1200 +
5 or 6 fish (harv)	$500–1000	$1000–2000	$2000 +

Table 7.3 continued

Type of violation	*First offence*	*Second*	*Third*
more than 6 fish harvested	\$600–2000	\$2000–4000	\$4000+
Using poisons, explosives or electrical charges	\$1000–2500 or damage assessment	\$2500–5000 or damage assessment	\$5000+ or damage assessment
Removal of or tampering with historical or cultural resources	\$500–5000	\$5000–10 000	\$10 000+
Discharge of substances	\$150–4000 or damage assessment	\$1000–8000 or damage assessment	\$2000+ or damage assessment
Damaging or tampering with markers	\$1000–2000	\$2000–4000	\$4000+
Dredging, filling or excavating activities	\$5000–10 000	\$5000–25 000	\$25 000+
Carrying or possessing firearms	\$100–500	\$500–1000	\$1000+
Use of bottom trawls or dredges	\$2000–5000 or damage assessment	\$5000–10 000 or damage assessment	\$10 000+ or damage assessment
Creating a wake in vicinity of divers or fishermen	\$100–500	\$500–1000	\$1000+
Use or discharge of explosives or weapons	\$250–1000 or damage assessment	\$1000–2000 or damage assessment	\$2000+ or damage assessment

resource damage cause by the grounding that was viewed along with the attenuating circumstances and the level of negligence surrounding the incident. The biological assessment had to provide the following information:

- A description of the habitat impacted.
- Dimensions and calculations of the total extent of area affected.
- Calculation of the extent of coral totally destroyed.
- Calculation of the extent of coral partially destroyed.
- Calculation of the proportion of living coral vs dead fossilized coral in the impacted area.

This information was used to calculate a total base penalty assessment depending on the type of habitat impacted and the extent of resource damage. (The biological and physical assessments for natural resource damage action may have a more detailed focus than those assessments done solely for the purpose of administrative penalty.) Damage to corals, for example, was given a higher base penalty than damage to sea-grasses. Also, total resource damage was given a higher value than partial resource damage. This was followed by a calculation of a final penalty assessment by multiplying the base assessment by a factor corresponding to the gravity of the circumstances which gave rise to the grounding. Negligence could be given a value ranging from one to five and determined from the sanctuary officer's investigation and report of the violation. Other criteria taken into account for a final penalty were whether or not there were prior violations and any aggravating or mitigating circumstances. These criteria could also be ranked and a numerical value applied according to the severity. An example of a formula that can be generated from these data is:

Area of habitat totally destroyed (m^2) × penalty level (\$ m^{-2}) for that habitat × negligence factor (1–5) × aggravating circumstances = total penalty for the infraction (\$)

This procedure only applied to civil actions of grounding cases that were prosecuted in the Administrative Court. Damage actions filed in the US Courts for resource loss, recovery, etc. were not calculated using this method.

7.3.6 Seizure authority

Sanctuary officers had the authority to board, search and inspect any vessel suspected of being used in the commission of a sanctuary violation. They also had the authority to seize any evidence suspected of being used in committing a violation and any sanctuary resources taken during the infraction. Property subject to seizure included spear-guns or spearing devices used in spear-fishing violations, tropical fish collecting equipment, wire fish-traps, or any other piece of gear used during the violation. Vessels could be seized under some circumstances.

The handling and disposition of the seized evidence varied considerably. All seizures were photographed and documented, with property receipts being issued to the violator. Living resources and small amounts of destroyed resources (e.g. speared fish) were returned to the water after being photographed. Large amounts of marketable catch (e.g. fish or lobster) were sold to a fish house and the proceeds held by the General Counsel's Office pending disposition of the case. Spear-guns and other related seizures were destroyed one year after the case was settled, if the case was settled in favour of the government (Green, 1989).

7.3.7 Use of recovered penalty money

In 1988 the US Congress authorized the use of recovered penalty (fine) money, response costs and damages in the management of the national marine sanctuaries. Assessment and response costs (e.g. to vessel groundings) and damages recovered in civil actions, or civil penalties collected for sanctuary violations, may be used:

- to restore, replace or acquire the equivalent of the sanctuary resources which were the subject of the action;
- to manage and improve the national marine sanctuary within which are located the sanctuary resources which were the subject of the action; and
- to manage and improve any other national marine sanctuary.

Sanctuary management programmes where these funds are directed primarily include education, research and resource restoration.

The enforcement procedures described in this section have been operating for some time. As specific regulations are established for the Florida Keys NMS these same procedures will more than likely be implemented, although some variation in the handling of vessel groundings will probably occur due to the vast size of the Florida Keys NMS.

7.4 LOOE KEY NATIONAL MARINE SANCTUARY: 10-YEAR DEVELOPMENT OF A LAW ENFORCEMENT PROGRAMME

The enforcement procedures used at Key Largo and Looe Key developed over time and it is important to recognize that there may be different needs at different phases of a management programme for a sanctuary. This is illustrated by looking at the development of the law enforcement programme at Looe Key NMS.

Looe Key NMS was designated in January 1981 and on-site management began in spring 1982, when a manager was hired and Florida Park Service Rangers started patrolling the reef. Word spread quickly through the local community that a Park Ranger, in a marked boat, was patrolling Looe Key Reef on a daily basis (Causey, personal observation). At that time not all of these Rangers had the authority to issue citations but their presence in the sanctuary, in a boat marked with official insignia and decals identifying them as 'Park Rangers', was enough to catch the attention of the locals. The designation of the sanctuary, and a commitment by NOAA to see it managed, quickly became a reality, especially in the minds of those who had opposed it becoming a sanctuary.

7.4.1 Phased enforcement approach

The enforcement of sanctuary regulations went through several phases in the first 4 years. Realizing that many visitors to the site during the early years were unlikely to know that it had become a protected area, the decision was made to have an initial verbal-warning phase for some infractions.

After a year and a half, the next phase of a combination of verbal warnings and written warnings was instituted. This approach did not apply to second-time violators, or to cases of intentional poaching by locals who were familiar with the area. After about 3 years, fewer verbal warnings were being given and more cases were being documented with written warnings. Also, at about this same time, sanctuary officers began writing more sanctuary violations for persistent illegal activities, such as spear-fishing and queen conch harvesting.

As the sanctuary approached its 4th and 5th years, the directive to the sanctuary officers was to initiate a more aggressive phase of issuing sanctuary citations and to use written warnings only when the officers' discretion made them lean towards a more lenient action. In the early part of this phase, the rule of thumb was that if sanctuary resources had been destroyed then the case was consistently handled with a stern enforcement action.

The rationale for giving a verbal warning as opposed to a more serious action depended on a variety of circumstances, most of which weighed the educational opportunities against the collection of a fine. This option should always remain available to those officers who demonstrate ability to use discretion and have the ability to capitalize on an educational opportunity. On high-use days, such as holidays, it was extremely important for the officers to know when to address a minor infraction with a verbal warning as opposed to writing a warning or sanctuary violation. The time required (approximately 20 minutes) to write a citation for a minor infraction (e.g. not hoisting a dive flag) had to be weighed against the potential oversight of a major violation or resource threat somewhere else on the reef.

This phased approach to increasing the level of enforcement severity is probably not applicable to all MPAs, particularly if locals soon learn how to take advantage of the situation. In the case of the Florida Keys, the locals did not take advantage of this approach (to our knowledge), probably because they were never aware of it. The plan can be tailored to meet the needs of the area and incorporate any enforcement strategies that will lead to greater compliance in a shorter time-frame. It can be developed over time but some enforcement plan should be considered for newly established MPAs from the outset.

7.4.2 Enforcement philosophy

The mere presence of a ranger in the sanctuary began serving as a deterrent to would-be violators and initiated the enforcement programme at Looe Key NMS. This management strategy and approach to enforcement was the initial phase of the kind of interpretive law enforcement programme that is practised today. Modelled somewhat after the philosophy of the National Park Service, or that of the Florida Park Service, the law enforcement programme for Looe Key NMS stressed interpretation and education as a means to gain compliance with sanctuary regulations. Since its establishment in 1916, the US National Park Service has used law enforcement as a means to protect park resources, persons and property. The National Park Service law enforcement programmes emphasize the need for education through enforcement contacts using the 'least stringent means to achieve compliance by the person contacted' (US Department of the Interior, 1989).

Interpretive law enforcement at Looe Key NMS emphasized that the sanctuary officers should capitalize on educational opportunities but not in lieu of enforcement actions. Often referred to as 'preventive enforcement', the objective was to reach the public with educational messages and literature before they had the opportunity to damage the sanctuary's resources. Increased public awareness and support is essential to achieve resource protection goals in the long term (Dobbins, 1983) and, in theory, this form or philosophy of resource protection can be accomplished by every member of staff of an MPA and not just the commissioned law enforcement officers.

7.4.3 Off-site interpretation

As the sanctuary boundary was 5.6 km from the nearest point of land and was approachable by boat from 360 degrees, it was a challenge to get information to visitors prior to their trip to Looe Key Reef. This was particularly true since there was no land-based facility inside the sanctuary. The variety of interpretive techniques which were used to make the public aware of the resource protection goals of the sanctuary prior to their visit included:

- Slide presentations on request by sanctuary staff at campgrounds, schools and various public meetings.
- Interpretive displays set up and manned by sanctuary staff at boat shows and diving shows.
- Brochures, posters and other literature distributed along a regular route to dive shops, marinas, boat rental companies, campgrounds, real estate offices, motels, etc.
- Outdoor interpretive signs with a map and regulations at all boat ramps and dive shops in the vicinity of Looe Key NMS.
- Sanctuary videos available to the dive shops, schools and general public at no cost.
- Frequent users to the sanctuary, including dive charter captains, education groups, dive shop owners, and so on, were invited to an annual meeting where topics pertinent to the management of the sanctuary were discussed.

7.4.4 On-site interpretation

A number of on-site techniques for reaching the public with interpretive messages to help 'preventive enforcement' were very effective at Looe Key NMS. One was for sanctuary officers to use a telescoping boat-hook with a clip to pass brochure packets to visitors as they arrived in the sanctuary (Figure 7.5). These boat-to-boat encounters gave the officers an opportunity to make direct interpretive contact with the visitors, while simultaneously allowing them to check the visitor's boat discretely for any signs of an infraction. As the sanctuary officers approached a visitor's boat there were often a few awkward moments when visitors were puzzled about being approach by a law enforcement officer. Because of this predictable reaction, it was important for the officer to take charge of the encounter quickly and set the visitors at ease with a friendly greeting. It must be stressed that this applied to interpretive encounters and not predetermined enforcement actions. The officer then passed over a packet of brochures and visitors had the opportunity to ask questions.

Generally speaking, once an MPA becomes well known to the public, the majority of those visiting the area are aware of its special significance and have no intention of violating rules. For that reason, patrolling officers could use the friendly interpretive approach without feeling that they were

Figure 7.5 Sanctuary officers passing out brochures during boat-to-boat contacts.

compromising their professional image as law enforcement officers. Regardless of the situation, however, the officers must remain in control of each encounter with visitors, and rely on their powers of observation to determine if a resource violation has occurred. If what started off as a routine interpretive encounter turns into a resource violation, then the use of discretion on the part of the officer can still result in a pleasant learning experience for the visitor.

During high-use holidays or busy weekends, volunteers were sometimes used to pass out brochures on-site. They used the same technique as the officers and tried to cover every boat arriving at the reef. This assistance freed officers to conduct their routine surveillance.

On-site interpretive snorkelling tours were frequently given by the sanctuary staff to special groups making such a request. Although the interpretive messages given during these tours covered many subjects, the regulations to help resource protection and enhancement were emphasized. These *in situ* interpretive tours were very helpful in informing the public about the need to protect the coral reef resources and the factors that influence the health of the reef.

Sanctuary officers periodically conducted underwater enforcement patrols. During these patrols the surveillance primarily focused on physical contact with the corals by divers and snorkellers. The officers carried a slate that identified them as sanctuary staff on one side and, on thc other side had a message to not touch the corals. The slate was flashed at the visitors or they could be signalled to surface depending on the severity of the infraction. Underwater patrols were an excellent way of directly monitoring the activities of the visitors and were carried out by staff other than law enforcement officers on occasions.

Additional enforcement support came from the regular users of the sanctuary such as the dive charter captains. They frequently witnessed infractions and served as an invaluable service by reporting their observations to the sanctuary officers. Looe Key NMS staff made it a policy to respond quickly to all the calls made by dive charter captains. In order to promote this kind of support, the sanctuary officers attempted to communicate frequently with captains and to keep them informed of new developments in the sanctuary.

7.4.5 Boundary buoys and mooring buoys

In addition to the on-site activities by sanctuary staff, construction projects helped to protect the resources of the sanctuary while simultaneously lessening the task of the sanctuary officers. Boundary buoys were installed in 1984 to mark the outer boundary and 'Core Area' of the sanctuary (Figure 7.6). These oriented visitors and also provided a means by which law-abiding

Figure 7.6 Marker buoys indicate sanctuary boundaries and Core Areas.

citizens could recognize when they were in the vicinity of the sanctuary if they were involved in an activity that was not allowed at Looe Key NMS.

Mooring buoys were in use at both sanctuaries in the Florida Keys (Halas, 1985) and although they were primarily recognized for their value in eliminating anchor damage to the corals, they also helped in other ways. Prior to the installation of mooring buoys at Looe Key NMS the majority of the officers' time during the beginning of a surveillance patrol was taken up by checking anchors to make sure they were not in coral. With buoys in place, an officer could, at a glance, make sure that none of the boats were anchored in coral.

7.5 CONCLUSIONS

There has been a considerable amount written in recent years about the management of MPAs but little attention has been focused on the enforcement programmes required to achieve the very basic goals of resource protection in these areas. The enforcement strategies for Key Largo and Looe Key NMSs (and to be continued in the Florida Keys NMS) which have been described in this chapter illustrate the issues that need to be addressed and explain the detailed working of an enforcement programme. This needs to be tailored to the needs of a specific site, in a specific programme, and in a setting that may be unique. However, the basic message is that immediate, planned enforcement should be an underlying premise for all MPAs, regardless of the setting.

ACKNOWLEDGEMENTS

Special thanks go to Ms Darlene Finch, Mr Frederick Kyle, Mr Dennis Riley and Ms Laura Goddard Causey for their critical review of this manuscript. Additional gratitude is expressed to Ms Joy Tatgenhorst and Ms Georgette Carver of the Florida Keys NMS staff for their assistance in preparing portions of this chapter.

It goes without saying that, if it were not for the hard work and dedication of the sanctuary officers at Key Largo NMS and Looe Keys NMS, the enforcement programmes would not have achieved so much. In recognition for a job well done, my sincerest gratitude goes to Sanctuary Officers Benny J. Davis, William B. Green, Rodney K. Hartsing, Scott E. Larosa, Jayson R. Horadam, Harry 'Beau' Jackson and Mary Taglarini. In the early days of the enforcement programme the efforts of Florida Park Service staff including Carl Neilsen, Ron Johns, George Jones, Henry T. Markey and Andrea Ash (among others) were also very much appreciated.

APPENDIX 7.A KEY LARGO NATIONAL MARINE SANCTUARY: FINAL REGULATIONS (15 CFR – PART 929)

§929.6 Allowed Activities

All activities except those specifically prohibited by §929.7 or other applicable authority may be undertaken within the Sanctuary subject to the restrictions and conditions imposed by other authorities.

§929.7 Activities Prohibited or Controlled

(a) Unless permitted by the Assistant Administrator in accordance with §929.10, or as may be necessary for the national defense, or to respond to an emergency threatening life, property or the environment, the following activities are prohibited or controlled within the Sanctuary. All prohibitions and controls must be applied consistently with international law. Refer to §929.9 for penalties for commission of prohibited acts.

(1) Removal or damage of natural features, marine life and archaeological and historical resources.

(i) No person shall destroy, injure, harmfully disturb, break, cut or similarly damage or remove any coral or other marine invertebrate, or any plant, soil, rock, or other material, except that commercial taking of spiny lobster and stone crab by trap and recreational taking of spiny lobster by hand which is consistent with both the applicable regulations under the appropriate Fishery Management Plan and these regulations is allowed. Divers are prohibited from handling coral formations, standing on coral formations, or otherwise disturbing the corals.

(ii) No person shall catch or collect any tropical fish.

(iii) No person shall remove, deface, damage, or tamper with archaeological or historical resources or the cargo of any submerged wreck or other historical resource within the boundaries of the Sanctuary.

(iv) There shall be a rebuttable presumption that any items listed in these paragraphs found in the possession of a person within the Sanctuary have been collected or removed from within the Sanctuary.

(2) Dredging, filling, excavating and building activities. No person shall dredge, excavate, fill or otherwise alter the seabed in any way nor construct any structure of any kind, whether permanent or temporary, with the exception of navigation aids.

(3) Discharges. No person shall deposit or discharge any materials or substances of any kind into the waters of the Sanctuary. The only exceptions are:

(i) Fish or fish parts and chumming materials;
(ii) Cooling waters from vessels; and
(iii) Effluent from marine sanitation devices approved by the United States Coast Guard.

(4) Tampering with markers. No person shall mark, deface or damage in any way whatsoever, or displace, remove or tamper with any signs, notices or placards, whether temporary or permanent, or with any navigational aids, monuments, stakes, posts, mooring buoys, scientific equipment or other boundary markers installed by the Sanctuary Manager, or trap floats placed for the purpose of lobster fishing.

(5) Use of harmful fishing methods. No person shall use within the Sanctuary, or shall carry or possess, except while passing without interruption through the Sanctuary or for law enforcement purposes, the following firearms or weapons: Pole spears, air rifles, bows and arrows, slings, Hawaiian slings, rubber powered arbaletes, pneumatic and spring loaded guns, explosive powered guns or similar devices known as spearguns. No person shall use within the Sanctuary:

(i) Wire fish traps;
(ii) Bottom trawls, dredges, fish sleds, or similar vessel-towed or anchored bottom fishing gear or net; or
(iii) Poisons, electric charges, explosives or similar devices.

(6) Operation of watercraft and anchoring. All watercraft shall be operated in accordance with applicable Federal rules and regulations. The following additional regulations apply within the boundaries of the Sanctuary.

(i) Watercraft shall be operated to avoid striking or otherwise causing damage to the natural features of the Sanctuary.

(ii) Watercraft must use mooring buoys, stations or anchoring arms when such facilities have been provided.

(iii) No anchor shall be cast or dragged in such a way as to damage any coral reef formations. Anchors shall be dropped only on sand flats off the reefs and be placed to avoid dragging into the coral formations.

(iv) Within 100 yards of divers, sightseeing boats or fishermen, no watercraft shall be operated at a speed greater than 4 knots or in any manner to create a wake, except by law enforcement officials while in the performance of their official duties.

(v) All watercraft from which diving operations are being conducted shall fly in a conspicuous manner the red and white 'divers down' flag. Divers shall stay within 100 yards of their diving flag.

(7) Use of dangerous weapons. Except for law enforcement purposes, no person shall use or discharge explosives or weapons of any description within the Sanctuary boundaries. Distress signaling devices, necessary and proper for safe vessel operation, and knives generally used by fishermen and swimmers are not considered weapons for purposes of this subsection.

(b) The Sanctuary may be closed to public use in the event of emergency conditions endangering life or property. The Assistant Administrator or his/her designee may also close certain areas in order to permit recovery of the living resources from overuse, or provide for scientific research relating to protection and management. However, the total closed area shall not exceed a size necessary to accomplish these purposes. Public notice of closures will be provide through the local news media and posting of placards at the John Pennekamp Coral Reef State Park, if deemed necessary.

(c) The regulation of activities within the Sanctuary shall not prohibit any activity conducted by the department of Defense that is essential for national defense or because of emergency. Such activities shall be conducted consistently with all regulations to the maximum extent possible.

(d) The prohibitions in this section are not based on any claim of territoriality and will be applied to foreign persons and vessels only in accordance with recognized principles of international law, including treaties, conventions and other international agreements to which the United States is signatory.

APPENDIX 7.B LOOE KEY NATIONAL MARINE SANCTUARY: FINAL REGULATIONS (15 CFR – PART 937)

§937.5 Allowed Activities

All activities except those specifically prohibited by §937.6 may be carried on in the Sanctuary subject to all prohibitions, restrictions and conditions imposed by any other authority.

§937.6 Activities Prohibited Without a Permit

(a) Unless permitted by the Assistant Administrator in accordance with §937.8, or as may be necessary for the national defense, in accordance with Article 5, Section 2 of the Designation, or to respond to an emergency threatening life, property or the environment, the following activities are prohibited within the Sanctuary. All prohibitions must be applied consistently with international law.

(1) Removing or damaging distinctive natural features.

(i) No person shall break, cut or similarly damage or take any coral or marine invertebrate except as an incidental result of anchoring outside the Fore Reef where sand anchoring is encouraged but not required. Divers are prohibited from handling coral or standing on coral formations.

(ii) No person shall take, except incidentally to allowed fishing activities, any tropical fish or marine invertebrate.

(iii) There shall be a rebuttable presumption that any items listed in this paragraph found in the possession of a person within the Sanctuary have been collected or removed from within the Sanctuary.

(2) Operation of watercraft. All watercraft shall be operated in accordance with Federal rules and regulations that would apply if there were no sanctuary. The following constraints also shall be imposed.

(i) No person shall place any anchor on coral within the Fore Reef of the Sanctuary nor allow any chain or rope to enter the Fore Reef in a way that injures any coral. When anchoring dive boats, the first diver down shall inspect the anchor to ensure that it is placed off the corals and will not shift in such a way as to damage corals. No further diving is permitted until the anchor is placed in accordance with these requirements.

(ii) Watercraft must use mooring buoys, stations or anchoring areas when such facilities and areas have been designated and are available.

(iii) Watercraft shall not be operated in such a manner as to strike or otherwise cause damage to the natural features of the Sanctuary.

(iv) All watercraft from which diving operations are being conducted shall fly in a conspicuous manner, the red and white 'divers down' flag.

(3) Using harmful fishing methods.

(i) No person shall use or place wire fish traps within the Sanctuary.

(ii) No person shall place lobster traps within the Fore Reef area of the Sanctuary.

(iii) No person shall use pole spears, Hawaiian slings, rubber-powered arbalets, pneumatic and spring loaded guns or similar devices known as spearguns within the Sanctuary.

(iv) No person shall use poisons, electric charges, explosives or similar methods within the Sanctuary.

(4) Removing or damaging distinctive historical or cultural resources. No person shall remove, damage or tamper with any historical or cultural resources, including cargo pertaining to submerged wrecks.

(5) Discharges. No person shall deposit or discharge any materials or substances of any kind except:

(i) Fish or parts and chumming materials.
(ii) Cooling water from vessels.
(iii) Effluents from marine sanitation devices.

(6) Markers.

(i) No person shall mark, deface or damage in any way or displace, remove or tamper with any signs, notices, or placards, whether temporary or permanent, or with any monuments, stakes, posts or other boundary markers installed by the managers or markers placed for the purpose of lobster pot fishing.

(ii) All activities currently carried out by the Department of Defense within the Sanctuary are essential for the national defense and, therefore, not subject to these provisions. The exemption of additional activities having significant impacts shall be determined in consultation between the Assistant Administrator and the Department of Defense.

(iii) The prohibitions in this section are not based on any claim of territoriality and will be applied to foreign persons and vessels only in accordance with principles of international law, including treaties, conventions and other international agreements to which the United States is signatory.

REFERENCES

Causey, B.D. (1989) Experiences from a National Marine Sanctuary in the United States. *Proceedings of the International Coastal Resources Management Workshop*, Belize, pp. 37–44.

Causey, B.D. (1990) Biological assessments of damage to coral reefs following physical impacts resulting from various sources, including boat and ship groundings. *Proceedings of the American Academy of Underwater Sciences Tenth Annual Scientific Diving Symposium*, St Petersburg, pp. 49–57.

Christhilf, F.D. (1986) Marine Sanctuaries, in *Environmental Quality*, Council on Environmental Quality 17th Annual Report, pp. 117–175.

Clark, J.R., Causey, B.D. and Bohnsack, J.A. (1989) Benefits from coral reef protection: Looe Key Reef, Florida. *Coastal Zone '89*, pp. 3076–3086.

Dobbins, J. (1983) *Looe Key National Marine Sanctuary Management Plan*, US Dept of Commerce, 58 pp.

Foster, N. and Lemay, M.H. (eds) (1989) *Managing Marine Protected Areas, An Action Plan*, Dept of State Publication 9673, 66 pp.

Green, W.B. (1989) *Enforcement at Key Largo and Looe Key National Marine Sanctuaries*, unpublished report, 4 pp.

Halas, J.C. (1985) A unique mooring system for reef management in the Key Largo National Marine Sanctuary. *Proceedings of the Fifth International Coral Reef Congress*, **4**, 237–242.

Hudson, B.E.T. (1988) Enforcement, in *Coral Reef Management Handbook* (eds R.A. Kenchington and B. Hudson), UNESCO, Jakarta, Indonesia, pp. 183–186.

Jaap, W.C. (1984) *The Ecology of the South Florida Coral Reefs: A Community Profile*, Mineral Management Service MMS 84-0038, 138 pp.

Miller, J.W. (ed.) (1988) *Results of a Workshop on Coral Reef Research and Management in the Florida Keys: A Blueprint for Action*, National Undersea Research Program Research Report 88-5, 49 pp.

OAS (1988) *Inventory of Caribbean Marine and Coastal Protected Areas*, Organization of American States, Dept of Reg. Development, Washington DC, 146 pp.

Office of Federal Register National Archives and Records Service General Service Administration (1985) *Code of Federal Regulations Commerce and Foreign Trade*, 15 Part 400 to end.

Salm, R.V. and Clark, J.R. (1984) *Marine Protected Areas: A Guide for Planners and Managers*, IUCN, Gland, Switzerland, 302 pp.

Sanctuaries Program Office (1982) *National Marine Sanctuary Program Development Plan*, 51 pp.

Tilmant, J.T. (1987) *Human Impacts of Recreational Activities on Coral Reefs: Facts and Recommendations* (ed. B. Salvat), Antenne Museum EPHE, French Polynesia, pp. 195–209.

US Dept of Commerce (1980) *Final Environmental Impact Statement, Proposed Looe Key National Marine Sanctuary*, 144 pp.

US Dept of the Interior (1989) *Law Enforcement Guideline NPS-9*, Release No. 3, National Park Service.

Wells, S.M. (ed.) (1988) *Coral Reefs of the World*, Vols 1, 2 and 3, IUCN Conservation Monitoring Centre, Cambridge, UK.

8

Caring for sea country – accommodating indigenous peoples' interests in marine protected areas

Dermot Smyth

8.1 INTRODUCTION

Over the last decade or so, there has been growing recognition by environmental conservation agencies of the special cultural, spiritual and economic relationship between indigenous peoples and the environment (IUCN *et al.*, 1991). In Australia, that recognition has translated into special management arrangements for some national parks which are now owned by Aboriginal people, who also control the management boards. There are currently four such parks in the Northern Territory (two of which are also listed as World Heritage areas) and negotiations are currently taking place for similar arrangements in other Australian states.

During this same period, a growing awareness of the need to protect and manage the marine environment around Australia has resulted in the establishment of several marine protected areas, notably the Great Barrier Reef Marine Park. To date, the accommodation of indigenous peoples' interests in marine protected areas has not yet developed to the same extent as in some terrestrial national parks.

This chapter examines the nature of the relationship between Australia's indigenous peoples and the marine environment and charts the still emerging efforts by marine park agencies to accommodate it, focusing on the Great Barrier Reef Marine Park (Figure 8.1) as a case study.

8.2 THE GREAT BARRIER REEF MARINE PARK

In 1975 the Australian Federal Government passed the Great Barrier Reef Marine Park Act, which created the Great Barrier Reef Region and provided for the Great Barrier Reef Marine Park (GBRMP) to be established in Commonwealth (i.e. federal government) waters within it. The GBRMP com-

Marine Protected Areas: principles and techniques for management. Edited by Susan Gubbay.
Published in 1995 by Chapman & Hall, London. ISBN 0 412 59450 1.

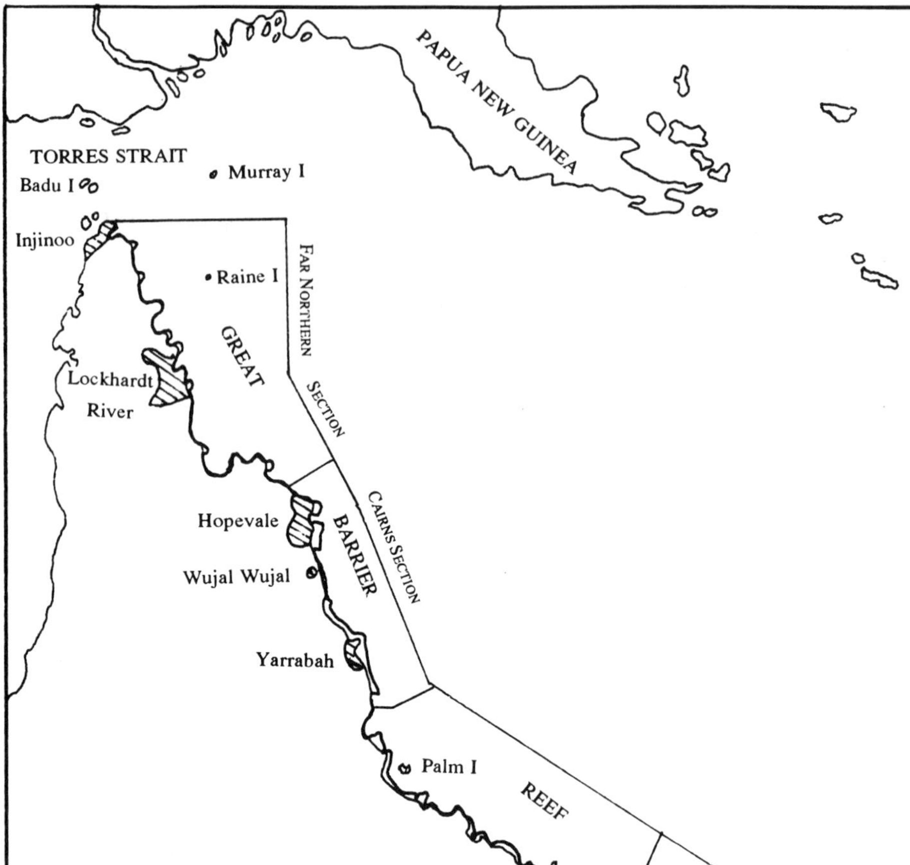

Figure 8.1 East coast of Queensland showing the Great Barrier Reef Marine Park, adjacent Aboriginal Trust Areas and the islands of the Torres Strait.

prises 98.5% of the Great Barrier Reef Region, the remainder being some coastal waters which were not considered, at the time, to be of sufficient conservation importance to the Great Barrier Reef to warrant inclusion in the GBRMP.

In 1981, the whole of the Great Barrier Reef as well as the islands within the Region were inscribed on the World Heritage List, under the UNESCO convention concerning the protection of the world's natural and cultural heritage.

During the 1980s the Queensland government established several state marine parks which include much of the coastal and intertidal waters adjacent to the GBRMP. Most of the islands throughout the barrier reef region are Queensland national parks.

The Great Barrier Reef World Heritage Area therefore contains three main conservation estates: the Great Barrier Reef Marine Park, coastal Queensland state marine parks, and island national parks. In practice there is a good deal of coordination in how these estates are managed. The federal government's Great Barrier Reef Marine Park Authority is responsible for the management of the GBRMP and for the development of general policy and zoning plans for the four sections of the GBRMP. Day-to-day management of the GBRMP is carried out by the Queensland government's Department of Environment and Heritage, which is also responsible for Queensland state marine parks and national parks.

The three-member Authority is advised by a Great Barrier Reef Marine Park Consultative Committee, comprising one member of the Authority and at least 12 other members, not less than one third of whom are nominees of the Queensland government. Committee membership is selected from a wide range of community, scientific and industry groups. Members are appointed on a personal basis; they are not delegates of their particular interest group.

The same members also constitute a parallel consultative committee which provides advice to the Queensland government on the management of Queensland state marine parks (but not island national parks).

The other major avenue for community involvement in the management of the marine parks within the Great Barrier Reef World Heritage Area is via a two-stage process of public participation in the development of zoning plans. The first occurs before the planning of a section of the marine park commences, to gauge the range of issues and concerns among members of the public. The second follows the preparation of a draft zoning plan.

Invitations to participate in these processes are communicated via advertisements in local newspapers, media releases and meetings arranged between marine park staff and interest groups in major centres along the Queensland coast.

This chapter primarily discusses issues relating to the involvement of indigenous peoples in the management of the GBRMP, though many issues discussed are also relevant to the management of adjacent Queensland state marine parks and island national parks.

8.3 INDIGENOUS PEOPLES OF THE MARINE PARK

There are two major indigenous populations with cultural, historical and economic interests in environments and resources contained within the GBRMP. These are Torres Strait Islanders and coastal Aboriginal people, both comprising many local groups with distinctive cultures, languages and maritime interests.

8.3.1 Torres Strait Islanders

Torres Strait Islanders are Melanesian people who have been occupying the islands of Torres Strait for at least the last 1000 years (Campbell, 1988). These islands formed about 6000 years ago when the land bridge between Australia and Papua New Guinea became flooded by the last great rise in sea level. Not surprisingly the sea and its resources are very important in the lives of the Islanders. Seafood consumption in Torres Strait has been ranked among the highest in the world (Johannes and MacFarlane, 1991). On Mabuiag Island in western Torres Strait, for example, seafood consumption was estimated to be 450 g per day, of which 60% was dugong meat, 29% turtle and 9% finfish (Nietschmann, 1989). Some Torres Strait Islanders have cultural and economic interests which extend southwards into the waters of the GBRMP.

Like all Torres Strait Island cultures, the Murray Islanders have a system of customary tenure which includes estates on both land and sea (Nietschmann, 1989; Johannes and MacFarlane, 1991). These estates are bounded areas containing many named places, some of which are sacred sites of spiritual significance related to mythological events surrounding the creation of land, sea and people. Within island estates, customary owners have rights to harvest wild food and plant gardens. Similarly, marine estate owners have customary rights to hunt and fish, and to allocate such rights to others.

The customary marine estates of Murray Islanders extend far beyond the visible horizon, south to Raine Island some 100 km south of Cape York, and well within the GBRMP. Customary ownership of these vast marine estates (which include Queensland, federal and international waters) is currently not legally recognized. Contemporary belief in customary ownership of marine estates, and right to resources within them, is based not only on tradition passed on through countless generations, but also on a history of commercial harvesting of pearl shell, top shells (*Trochus* species) and bêche-de-mer (holothurians) over the last 100 years.

People from throughout the Torres Strait have been involved in pearling and other commercial marine activities, variously run by Japanese and European entrepreneurs, missionaries and government agents since the late nineteenth century. During the middle decades of the twentieth century Badu became the most prosperous of the Torres Strait Islands, partially based on the utilization of resources within what is now the GBRMP.

Although the pearl shell industry is now confined to collecting live shell for use in cultured pearl farms, the collection of Trochus shell (used for making buttons) is still commercially viable. Torres Strait Islanders involved in the trochus industry have objected to restrictions placed on commercial fishing within large conservation zones of the far northern section of the GBRMP and the lack of consultation prior to the implementation of these restrictions (Smyth, 1993). The Marine Park Authority, however, contends that there was substantial consultation.

8.3.2 Coastal Aboriginal people

Aboriginal people have occupied the Australian land mass for at least 40 000 years and perhaps as much as 100 000 years. During that time Australia's coastline changed dramatically as sea levels fell and rose. The last major sea-level rise began about 20 000 years ago and stabilized between 6000 and 8000 years ago. This rise of over 100 m in sea level inevitably resulted in huge losses of Aboriginal territory as the wide continental shelf became flooded. While there is some evidence of a lapse in time between sea-level stabilization and Aboriginal occupation of particular parts of the coast (Beaton, 1985), it is reasonable to assume that Aboriginal people have been an integral part of maritime ecosystems since the establishment of the current Australian coastline.

Anthropological research over the last 60 years indicates the importance of the marine environment to coastal Aboriginal groups in north Queensland (e.g. Chase, 1980; Chase and Sutton, 1981; Hale and Tindale, 1933, 1944; Rigsby and Williams, 1991; Thompson, 1934). In traditional coastal societies, owned clan estates comprise portions of coastal land as well as extensive areas of sea. These integrated, bounded land/sea estates are owned by members of a clan descent group (usually through the male line) who traditionally have control of resource rights, as well as responsibility for the conduct of ceremonies necessary for the well-being of particular places or species.

For further information on customary marine tenure see Cordell (1989, 1991) and Bergin (1991).

Research commissioned by the Authority in recent years (e.g. Smith, 1987; Smyth, 1989, 1992), as well as statements by coastal Aboriginal people themselves (in Smyth, 1993; Wallace *et al.*, 1992), indicate that this close identification with maritime environments and resources persists in contemporary Aboriginal communities adjacent to the GBRMP. For example:

- Marine animals such as dugong, turtle, and stingray continue to have totemic importance for particular groups of people.
- Sacred sites which link people, animals, plants, landscapes and seascapes with the creation period of the Dreamtime continue to have profound meaning for coastal Aboriginal people (Figure 8.2).
- Subsistence hunting and fishing are widely practised, both as an economic activity and as a statement of belonging to and responsibility for customary marine estates and resources.

The extent to which these contemporary cultural interests in marine environments and resources persist varies from place to place along the Queensland coast, largely dependent on local history. In southern Queensland, and in some areas of the north, maritime Aboriginal culture has been very greatly disturbed by almost 200 years of European settlement. Over the last 100 years many Aboriginal people have been forcibly relocated to places far removed from their traditional estates. Until the 1960s it was common government practice to take Aboriginal children away from their families to be brought up out of contact with their own language and ways of life. Such events led to immeasurable personal suffering as well as considerable

Figure 8.2 Turtle Story place.

cultural loss between generations. In some places entire clans and language groups were killed or have been displaced without trace.

In spite of these grave pressures, very many Aboriginal people along the coast are knowledgeable about their customary estates, stories and sacred places. A common link with the past is the largely unbroken practice of subsistence hunting and fishing.

In far north Queensland, several Trust Area communities (former Aboriginal mission stations which are now largely managed by local people) are located on or close to customary estates. In such places, languages have survived and clan identification is strong. There is also considerable knowledge about the traditional ownership of maritime estates, the location of land and sea sacred sites, the conduct of ceremonies and behaviour taboos associated with particular places, species or the status of individuals in Aboriginal society.

8.4 CONTEMPORARY ABORIGINAL AND ISLANDER INTERESTS IN THE GREAT BARRIER REEF MARINE PARK

Contemporary interests of indigenous peoples in the management of the Great Barrier Reef Marine Park can be divided into cultural, economic and legal issues, all of which are interrelated.

8.4.1 Cultural interests

Cultural interests include the protection of sacred sites with the GBRMP, as well as the conduct of ceremonies associated with the well-being of particular places and species. Utilization of marine estates and resources provides opportunities for displaying, teaching and learning cultural skills, and knowledge relating to hunting and fishing. These activities also encourage the maintenance of language and interest in cultural traditions.

8.4.2 Economic interests

Economic interests fall into three main categories: subsistence hunting and fishing; commercial exploitation of marine resources; and vocational opportunities.

Subsistence hunting and fishing are activities which link coastal Aboriginal people with each other and through time with their ancestors. While serving important cultural functions, subsistence hunting and fishing continue to contribute significantly to household economies of Islander and Aboriginal people, including those living in coastal towns and cities.

Although Islanders resident in the Torres Strait may not often engage in subsistence hunting and fishing within the GBRMP, there are now sizeable

Torres Strait Islander populations located in major coastal settlements in north Queensland who wish to continue to have access to traditional marine resources, even though they have moved far away from their customary estates.

As mentioned above, Torres Strait Islanders have a long history of commercial exploitation of marine resources. Within the GBRMP, Islanders continue to be involved in Trochus fishing under licence from the Queensland government.

Although many coastal Aboriginal people were involved in the pearling, *Trochus* and bêche-de-mer industries in the past (usually as indentured divers but sometimes as skippers), these activities are no longer pursued. Coastal Aboriginal people are, however, expressing interest in establishing mariculture enterprises (particularly clam and oyster farms) and in small-scale commercial lobster fishing. A pilot oyster farm, partly owned by Aboriginal people, has been established within the Queensland Marine Park off Fitzroy Island near Cairns.

Over the last decade, coastal Aboriginal people have been establishing 'outstations' away from the large government settlements. Outstations are semi-permanent camps located at focal points of clan estates, where clan members and their relations can spend part of the year away from the major Aboriginal settlements. This trend, driven by social and cultural forces, has raised interest in exploring options for economic development in these small satellite communities. In such places, subsistence fishing and hunting are primary economic activities but small-scale commercial lobster fishing and mariculture are also viewed as possibilities.

Also viewed with interest is the possibility of gaining a livelihood from involvement in the management of marine parks and adjacent terrestrial environments. All coastal Aboriginal communities north of Townsville already employ their own Community Rangers to protect cultural and natural heritage and to manage tourism in their areas. Increasingly these rangers are working with government wildlife, marine park, customs and quarantine officers to provide unofficial coastal management in remote areas. At present, such cooperation is on an honorary basis, but coastal Aboriginal communities have expressed interest in undertaking formal management roles for which payment would be received from government agencies.

8.4.3 Legal interests

Australia's indigenous peoples have long pressed governments for legal recognition of their customary ownership of land and sea and their rights to resources. In 1991 the Queensland government passed the Aboriginal Land Act and the Torres Strait Islander Land Act, which established processes to enable some traditional lands to be returned to the appropriate indigenous peoples. In particular, some Crown land which is currently vacant, some national parks and all Aboriginal and Islander reserves and Trust Areas (former reserves) will be eligible for return to traditional owners, subject to approval by an Aboriginal Land Tribunal. The Land Acts also provide for the possibility of intertidal land to be returned to Aboriginal and Islander ownership.

An important limitation of the Acts is that the Queensland government retains control over which pieces of land will become available for claim. Particular areas of land must first be declared as available for claim before any Aboriginal or Torres Strait Island group can begin the claim process. To date the government has chosen to make available for claim only former Aboriginal reserves, nine national parks and some vacant Crown land in the far north of the state.

By mid 1994 several former Aboriginal reserves had been transferred to Aboriginal ownership. Aboriginal claims over three coastal national parks (Flinders Island, Cape Melville and Lakefield) and the Simpson Desert National Park were well advanced and the process of developing plans of management and appointing Aboriginal trustees could be expected to take place in 1995 (Figure 8.3). As yet no intertidal land has been made available for claim, but this may occur once the claims for coastal national parks have been resolved.

Figure 8.3 Aboriginal community rangers with patrol vessel.

While not providing for indigenous ownership of sub-tidal marine estates, these Acts are likely to alter radically the relationship between indigenous peoples and marine park management agencies. Firstly, the land claim process will clarify and identify who are the traditional owners for particular tracts of coastline, making it clear to conservation agencies which groups should be involved in the management of particular portions of marine parks. Secondly, Aboriginal ownership may be granted to intertidal land lying within marine parks. Thirdly, many islands within the GBRMP may be owned by Aboriginal people who could expect a major say in the management of surrounding waters.

Indigenous peoples resident on Trust Areas can obtain permits to carry out subsistence hunting and fishing within most of the GBRMP. Queensland fisheries legislation currently prevents indigenous peoples living outside Trust Areas from obtaining permits, though this is likely to change under provisions in the Nature Conservation Act 1992, once management plans have been prepared for individual species such as turtle and dugong.

Subsistence hunting and fishing are not recognized as legal rights of indigenous peoples within the GBRMP, though this has been advocated by the federal government's Law Reform Commission (Law Reform Commission, 1986). Similarly, Aboriginal and Islander rights to other marine resources, such as commercial fisheries and sea-bed minerals, are currently not legally recognized. These issues are considered further in the following section, which deals with a landmark decision of the Australian high court.

8.5 THE MABO CASE

In June 1992 a decision by the high court of Australia formally recognized the existence of Aboriginal 'native title' to land in Australia. Prior to this decision Australian governments had assumed that Aboriginal ownership of territory had been extinguished on the acquisition of sovereignty by the British in 1770. Similarly it had been assumed that indigenous ownership of the Torres Strait Islands was extinguished at the time of their formal annexation about 100 years later.

The case was brought before the Court by Eddie Mabo and other traditional owners of the Murray Islands in eastern Torres Strait. In 1982 the Islanders had refused an offer by the Queensland government to grant them a lease over their islands. The Islanders maintained that the government was not in a position to make such an offer since the Islanders believed that they already owned the islands under customary law. After a 10-year legal struggle, during which the Queensland government unsuccessfully attempted to extinguish native title to the islands by legislation[1], the Islanders' customary ownership was finally recognized as valid under Australian common law.

The successful claim strictly relates only to the land of one island within the Murray Group, but it has implications everywhere in Australia where indigenous peoples have retained customary association with their traditional country. Although the judgement deals specifically with land ownership, many legal commentators have suggested that there are implications for ownership of marine estates and rights to marine resources. One academic lawyer has concluded:

> Common law recognition of customary marine tenure is highly likely following the Mabo decision . . . there is nothing in the judgement which would seem to preclude the application of native title principles to the seabed. (Sutherland 1992)

Following a period of prolonged and intense negotiation between indigenous groups, federal and state governments, mining, farming, fishing and other interests, the Australian parliament passed the Native Title Act 1993. This Act provides legislative recognition of the Mabo high court decision and sets out procedures for the determination and protection of native title wherever it continues to exist. The legislation also provides mechanisms for claiming compensation in the case of customary landowners whose native title was wrongfully extinguished.

The Native Title Act 1993 provides for the possibility of native title being recognized in marine environments but the nature of such title will be a matter of determination by courts, probably in a series of test cases over the next few years. While it is now widely recognized by governments, conservation agencies and industry groups that native title continues to exist in coastal waters around much of Australia, especially in the north, there is uncertainty as to whether such title amounts to a property interest (as on land) or merely some form of access and use right (such as hunting, fishing and gathering rights). Whatever the precise outcome, the Mabo decision is already affecting the decision-making process of governments and the ways in which government agencies and indigenous groups interact. The possibility exists that legislation, regulations or administrative decisions which seek to curtail or extinguish native title or rights may now be subject to injunction or claims for compensation. In general terms, the Mabo decision has raised the prospect that agencies such as the Marine Park Authority may have to deal with indigenous peoples as 'owner groups' rather than 'user groups'.

---

[1] In 1985 the Queensland parliament passed the Queensland Coastal Islands Declaratory Act, the function of which was to extinguish any outstanding indigenous title or claim to the islands of Torres Strait. In a separate high court action in 1988 the Murray Islanders established that this legislation contravened provisions of the Racial Discrimination Act 1975. The Queensland Coastal Islands Declaratory Act was therefore declared invalid and the original claim for recognition of traditional ownership of the Murray Islands proceeded to the high court.

A special challenge for the Marine Park Authority will be the development of mechanisms to evaluate competing interests by indigenous groups and reef-based tourism operators. At present the Authority's assessment staff consult with appropriate Aboriginal people whenever an application is received for permission to conduct some new tourism venture. Such ventures often involve bringing many hundreds of people (sometimes more than 1000) to a reef each day and the long-term mooring of large pontoons. In areas adjacent to Aboriginal communities such high levels of tourism activities effectively prevent the exercise of Aboriginal hunting and fishing rights, and hence adversely impact on the maintenance of Aboriginal cultural and heritage values.

While the Authority's own regulations require them to protect Aboriginal cultural and heritage values, there is currently no direct participation by Aboriginal groups in the final decision on whether or not to grant a permit for such tourism ventures. That is, while Aboriginal groups are adequately consulted about such proposals, it is the Authority which makes the final judgement about which of the competing interests will prevail. Sometimes direct negotiations between tourism operators and Aboriginal groups result in modifications to the proposals to the satisfaction of all parties. However, where operators refuse to enter into such negotiations, Aboriginal people remain dependent on the judgement of the Authority in relation to the protection of cultural and heritage values.

8.6 ABORIGINAL AND ISLANDER INVOLVEMENT IN THE GBRMP

Current opportunities for Aboriginal and Torres Strait Islander involvement in the management of the GBRMP include:

- Membership of the Great Barrier Reef Marine Park Consultative Committee.
- Public participation process for the development of zoning plans.
- Attendance at marine park workshops and conferences.
- Participation in research projects.
- Participation in community liaison meetings.
- Employment by the Authority and the Queensland Department of Environment and Heritage.

8.6.1 Membership of the Consultative Committee

It was not until 1988, 13 years after the establishment of the GBRMP, that the first Aboriginal person was appointed to the Consultative Committee. In earlier committees, the interests of indigenous peoples were represented by the administrative head of the Queensland government department responsible for Aboriginal and Islander affairs.

In 1988 the Aboriginal person appointed to the Consultative Committee was at the time chairman of the Aboriginal Coordinating Council (ACC), an organization established by the Queensland government to provide advice on matters relating to the management of Queensland's 16 designated Aboriginal Trust Areas. Six of these communities have responsibilities for managing land adjacent to the GBRMP, and substantial cultural and economic interests within the marine park. The appointment was made with little prior consultation with coastal Aboriginal communities and the person selected attended no meetings of the Consultative Committee during his 2-year term.

In 1990, the ACC nominated one of its employees as the replacement member of the Consultative Committee. This person made a significant contribution to the deliberations of the Committee and in 1991 prepared a report (Ziegelbauer, 1991) which was critical of the Authority's progress in involving Aboriginal people in management of the GBRMP.

8.6.2 Public participation in the planning process

In theory, Aboriginal and Torres Strait Islander people have the same access to the public participatory planning process as other Australians interested in the future of the GBRMP. In practice, however, the style, pace and format of the planning process has largely excluded indigenous peoples, especially those with detailed cultural knowledge of marine estates within the marine park. For many such people, English is a second or even a third language, and they may have infrequent access to newspapers in which invitations to participate in the planning process are placed. Furthermore, such people are often not in a position to make detailed written submissions outlining their concerns.

Although officers of the marine park did visit the main coastal Aboriginal communities during the development of the first zoning plans, it appears that these plans were developed with only limited input from indigenous peoples. Aboriginal and Islander people did, however, attend a marine park workshop in 1978 prior to the development of the first far northern section zoning plan.

In 1988 the Authority funded a project to assist two Aboriginal communities to prepare submissions in relation to the second zoning plan for the Cairns section of the GBRMP (Smyth, 1989). Those submissions, based on meetings held with traditional owners of marine estates and the general Aboriginal community, outlined Aboriginal aspirations for greater involvement in the management of the marine park and especially the protection of their marine sacred sites. The Hopevale Aboriginal Community outlined similar concerns in a submission to the Authority as part of the Cairns section re-zoning process (Hopevale Community Council, 1990).

8.6.3 Attendance at marine park workshops and conferences

Coastal Aboriginal and Torres Strait Islander people have participated in several workshops and conferences organized by the Marine Park Authority which focused on indigenous interests in the marine environment. These include:

- 1978: Far Northern Section Planning Workshop.
- 1985: Traditional Knowledge of the Marine Environment Workshop (Bani, 1988; Mabo, 1988).
- 1988: Innovative Planning and Management Workshop (Pearson, 1988).
- 1990: Torres Strait Baseline Study Conference (Lawrence and Cansfield-Smith, 1991).

These events have given selected Aboriginal and Islander people the opportunity to put on record the broad range of issues which characterize their relationship with the marine environment. Also participating in these workshops and conferences were anthropologists, archaeologists and ethno-ecologists with research interests in coastal Queensland and Torres Strait. Papers delivered by these specialists provided further evidence of the complexity and continuity of Aboriginal and maritime cultural and economic interests (e.g. Chase, 1981).

8.6.4 Participation in research projects

Indigenous peoples have been involved in several research projects funded by the Marine Park Authority. These include:

- A study on the usage or marine resources by Aboriginal communities on the east coast of Cape York Peninsula (Smith, 1989).
- Studies on Aboriginal maritime culture in the Cairns and far northern sections of the GBRMP (Smyth, 1990, 1992).
- Studies on dugong populations (Marsh *et al.*, 1984).
- Studies on the usage of marine resources by Torres Strait Islanders, as part of the Torres Strait Baseline study (Lawrence and Cansfield-Smith, 1991).

These projects have provided selected indigenous peoples with opportunities to participate in, and contribute to, research relating to the management of the GBRMP. However, indigenous peoples have so far had only limited involvement in the selection and design of research projects and currently have no meaningful involvement in how the results are utilized in management.

8.6.5 Participation in community liaison meetings

Officers of the Marine Park Authority and the Queensland Department of Environment and Heritage have visited coastal Aboriginal communities from time to time since the establishment of the GBRMP. Such meetings have provided opportunities for traditional owners of marine estates, custodians of sacred sites and people actively engaged in subsistence hunting and fishing to communicate directly with marine park personnel. The Authority and Department have not been sufficiently well resourced to conduct such meetings regularly. As a result, liaison meetings have generally been arranged in response to particular problems rather than to negotiate and develop cooperative management arrangements.

Events which led to such meetings include the decision by the Authority to impose a quota on dugongs caught by the Hopevale Aboriginal Community in 1993 (Marsh *et al.*, 1984) and the Authority's decision to permit tourist boat operations in Northeast Bay on Palm Island, an Aboriginal Trust Area (Smyth, 1993). In both cases, liaison meetings resulted in the reversal of the original decision, pointing both to the success of such on-ground liaison and the need for it to be on-going and proactive, rather than ad hoc and reactive.

The Marine Park Authority has been taking a more proactive approach since the appointment of an Aboriginal Liaison Officer in 1992. A similar approach is being taken by the Department of Environment and Heritage, which has entered into negotiations with the Yarrabah Aboriginal community over the development of cooperative management arrangements for Mission Bay, which is part of the Cairns state marine park adjacent to that community. A draft management plan for Mission Bay has been completed which will, if implemented, lead to substantial devolution of management to the community rangers employed by the Yarrabah Aboriginal Council.

8.7 ASSESSMENT OF ABORIGINAL AND ISLANDER INVOLVEMENT

Notwithstanding the potential implications of the recent Mabo high court decision, it cannot yet be said that indigenous peoples' interests are adequately accommodated in the management of the GBRMP. Contributing factors to this situation may include:

- The still emerging understanding by the Authority of the scope of Aboriginal and Torres Strait Islanders in the marine park.
- Political considerations in achieving cooperation between the Australian and Queensland Governments in the management of the GBRMP.
- A belief among the general Australian community that making special

provisions for indigenous peoples runs contrary to democratic ideals of equality.

- Lack of legislative recognition of Aboriginal and Islander interests.

Each of these issues is discussed in the following subsections.

8.7.1 Understanding Aboriginal and Islander interests

From the beginning, the Marine Park Authority has accepted the legitimacy of Aboriginal and Islander interests in subsistence hunting and fishing, which have been accommodated within the limitations imposed by Queensland fisheries legislation. Only recently, however, has the Authority begun to take steps to accommodate the wider cultural and economic relationship between indigenous peoples and the marine environment, which legitimizes a more substantive role in management.

It may initially be difficult to understand why this broader understanding of indigenous peoples' interests in the marine environment has taken so long to develop. Through workshops, conferences, liaison meetings and research projects, indigenous people have made it clear that hunting and fishing are part of a complex cultural relationship between people and environment. They have also made it clear that they wish to regain some control over their customary marine estates.

That the Marine Park Authority has been slow to respond to the full extent of indigenous peoples' interests may be due in part to the cultural values held by the non-indigenous Australian community (including Authority members and staff) which differ markedly from Aboriginal and Islander values with respect to the marine environment. Aboriginal and Islander concepts of integrated land/sea estates, ownership and stewardship of bounded areas of water, sea-bed sacred sites and exclusive clan or family rights to particular marine resources are not easily accommodated into the European view of oceans as global, national or community commons.

In the decade and a half since the establishment of the GBRMP, there has been a gradual awakening of interest and understanding among the wider community about Aboriginal and Torres Strait Islander issues. The Authority has responded gradually to that awakening and has itself contributed to the process by conducting research into Aboriginal maritime interests.

8.7.2 The politics of intergovernmental cooperation

Differing world views aside, the politics of maintaining cooperative relations between the Marine Park Authority (a federal government agency) and the

Queensland government may also have restricted opportunities to accommodate Aboriginal and Islander interests more widely.

For most of the history of the GBRMP, the federal government has been more predisposed to recognition of indigenous peoples rights than the Queensland government. It is possible that, had the Marine Park Authority substantially recognized Aboriginal and Islander interests in maritime ownership and management, it would have jeopardized other aspects of marine park management which required agreement with the Queensland government.

The election in 1989 of a new Queensland government which was committed to greater recognition of Aboriginal and Torres Strait Islander rights, has enabled the Authority to explore the possibility of more comprehensive accommodation of indigenous peoples' interests in the marine park. The passage of the Aboriginal Land Act 1991 by the Queensland parliament has focused attention on the relationship between Aboriginal people and land, with flow-on implications for their interests in the sea. The Mabo high court decision has provided further impetus to examining the full relationship between indigenous peoples and the Australian environment.

8.7.3 Discrimination by equality

Reinforcing the view of the sea as a common domain is the widespread belief that recognition of special rights for indigenous peoples runs contrary to democratic ideals of equality for all. Such beliefs have not only restricted recognition of Aboriginal and Islander traditional rights, but have also discriminated against their access to processes of public participation in marine park planning. Use of printed standard English in planning documents, communicating via advertisements in newspapers and allowing inadequate time for community consultations, have all led to the exclusion of Aboriginal and Islander input into the planning process.

Similarly, the failure to appoint an Aboriginal or Islander person to the Consultative Committee for the first 13 years of its operation may have been defended on the basis that no other members were appointed on the basis of race or ethnicity. The result, however, was to deny structural input by indigenous peoples into planning and management during the formative years of GBRMP operations.

This does not imply a wilful desire on the part of the Authority to exclude Aboriginal or Islander input. Rather it reflects the prevailing operations of Australian governments and government agencies during the 1970s and 1980s. During this period the level of communication and articulation between indigenous and non-indigenous Australians was generally very poor.

8.7.4 Lack of legislative recognition of Aboriginal and Islander interests

A further and related barrier to broad Aboriginal and Islander interests in the GBRMP is the current lack of reference to such interests in the Great Barrier Reef Marine Park Act. Similarly, the Great Barrier Reef World Heritage Area was nominated on the basis of its 'natural' values alone (GBRMPA, 1981). While the nomination document referred to earlier Aboriginal maritime societies and the presence of coastal and island archaeological sites, contemporary Aboriginal interests are described only with reference to subsistence hunting and fishing.

Similarly, the legislative requirement to establish a Consultative Committee does not specify that Aboriginal and Islander people should be represented. Without such specific reference in legislation, social and political pressures which ignore indigenous peoples' interests can prevail.

Furthermore, the absence of a legislative requirement to accommodate Aboriginal and Islander interests may limit the extent to which the Authority can obtain government support to devote adequate resources to resolving these issues, for example funding to employ adequate liaison staff.

8.8 RESOLVING ABORIGINAL AND ISLANDER INTERESTS IN THE GBRMP

The Mabo judgement came at a time when the Authority was already contemplating major reforms in the ways in which it accommodated indigenous peoples' interests in the marine park. In cooperation with other government agencies and community groups, the Authority has developed a 25-year strategic plan for the Great Barrier Reef World Heritage Area (GBRMPA, 1994). The strategic plan, which has been endorsed by the federal and Queensland governments, outlines specific objectives and strategies with respect to Aboriginal and Islander interests over a 25-year period. The strategic plan also outlines objectives and strategies to be accomplished over the first 5-year period. These are set out in Table 8.1.

These provisions for accommodating indigenous peoples' interests in the World Heritage Area (including the GBRMP) were developed prior to the high court decision in the Mabo case. Subsequent to that decision, the Cape York Land Council (which had represented Aboriginal interests during the development of the plan) announced that they would not endorse the plan on the grounds that it did not adequately recognize the full range of Aboriginal maritime interests that may now exist in common law as a result of the Mabo judgement.

Notwithstanding the native title implications which may not be addressed in the strategic plan, it does contain provisions which could significantly improve Aboriginal control over customary marine estates with the GBRMP. Most important of these are Objective 6.4 and Strategies 6.4.1, 6.4.2 and 6.4.3, which provide for the development of cooperative management arrangements with indigenous peoples for specific areas of the marine park, as well as the legislative basis to establish such arrangements.

Other objectives and strategies dealing with the development of culturally relevant interpretive material, membership of advisory boards and the development of Aboriginal and Islander training and employment programmes are also important initiatives. In the long run they are likely to be successful only if they are developed as adjuncts to the central issue of devolving control of customary marine estates within the marine park to appropriate indigenous traditional owners.

Several options for achieving greater control over customary maritime estates by appropriate indigenous peoples have been proposed to the Authority in several reports (e.g. Baldwin, 1984; Smith, 1989; Smyth, 1990, 1992; Ziegelbauer, 1991). These options include:

- The establishment of a separate Aboriginal and Torres Strait Islander Consultative Committee to advise the Authority on planning and management issues relevant to indigenous peoples' interests.
- The establishment of Aboriginal-controlled management boards for sections of the marine park.
- The establishment of local Aboriginal and Islander management committees to be responsible for designated areas within the marine park where indigenous interests are recognized as paramount.

These options are neither exhaustive nor mutually exclusive, and it is likely that different solutions would be appropriate for different regions of the GBRMP. For any solutions to be workable, they will have to emerge after a process of consultation and negotiation with maritime Aboriginal and Islander groups at a local level, as well as with their representative organizations.

During 1994, as part of the zoning review of the Far Northern Section of the Great Barrier Reef Marine Park, the Authority has commenced such negotiations; the development of the new zoning plan for this section (and associated management arrangements) is likely to be a watershed in relations between the coastal indigenous peoples and the Authority. The zoning review provides an opportunity to explore mechanisms for devolving at least some management responsibilities to local or regional indigenous organizations to reflect the reality of their continued cultural and economic association with the marine environment. Meanwhile the Commonwealth Government is giving consideration to amending the Great Barrier Reef Marine Park Act to recognize formally indigenous hunting and other interests, and to guarantee membership of indigenous peoples on the Community Consultative Committee and on the Authority itself.

Table 8.1 Strategic plan for the Great Barrier Reef World Heritage Area: Aboriginal and Islander interests

Objective 6.1:
To ensure that the interests of Aborigines and Torres Strait Islanders are reflected in the management of the Area.

Strategies
6.1.1: Develop effective participation processes and structures in conjunctions with Aborigines and Torres Strait Islanders;
6.1.2: Cooperatively develop guidelines for stakeholder agencies and organizations for culturally appropriate interaction with Aborigines and Torres Strait Islanders;
6.1.3: Ensure negotiations occur on all aspects of management;
6.1.4: Ensure that use by Aborigines and Torres Strait Islanders is taken into account in the development of resource management plans;
6.1.5: Where plans are appropriate, Aborigines and Torres Strait Islanders to develop, with stakeholder agencies and organizations, management plans to ensure that their traditional use of resources is ecologically sustainable;
6.1.6: Provide opportunities for Aborigines and Torres Strait Islanders for membership on, and full involvement in, advisory committees and management boards;
6.1.7: Develop and implement employment and training programs in stakeholder organizations of Aborigines and Torres Strait Islanders;
6.1.8: Consider the implications of relevant legislation for native title.

Objective 6.2:
To inform the general public of the cultures and economies of Aborigines and Torres Strait Islanders in relation to the Area.

Strategies
6.2.1: Develop educational and interpretive materials and programs, in conjunction with Aborigines and Torres Strait Islanders;
6.2.2: Incorporate information about Aborigines and Torres Strait Islanders in education curricula and interpretive programs.

Objective 6.3:
To develop a culturally appropriate information program for Aborigines and Torres Strait Islanders regarding the Area and its management.

Strategies
6.3.1: Produce culturally appropriate material;
6.3.2: Disseminate information in a culturally appropriate manner.

Objective 6.4:
To establish cooperative management arrangements between Aboriginals and Torres Strait Islanders and stakeholder agencies in the area.

Strategies
6.4.1: Establish a legislative basis for cooperative management arrangements;

Table 8.1 continued

6.4.2: Establish cooperative management arrangements for specific areas;
6.4.3: Provide for Aboriginal and Torres Strait Islander representation on advisory committees and management boards.

Objective 6.5:
To ensure that projects relating to social, cultural and economic interests of Aborigines and Torres Strait Islanders are included in research and monitoring programs.

Strategies
6.5.1: Identify and develop relevant research and monitoring projects in consultation with Aborigines and Torres Strait Islanders;
6.5.2: Involve Aborigines and Torres Strait Islanders in projects that affect the interests of their people.

Several initiatives by the Marine Park Authority over the last two years indicate that the process of accommodating the special interests of indigenous peoples in the marine park is gathering momentum. These initiatives include:

- The appointment of an Aboriginal Liaison Officer.
- The development of an Aboriginal employment strategy.
- The development of a marine park training programme for Aboriginal Community Rangers.
- The adoption of a Regulation which amends the assessment criteria for the development of zoning plans so as to protect the cultural and heritage values held in the marine park by traditional inhabitants and other people.
- The establishment of Aboriginal Councils of Elders at several locations along the Queensland coast to assist the Authority in determining the allocation of dugong and turtle hunting permits[2].
- The training and employment of two Aboriginal Community Rangers by the Authority at each of three coastal communities to assist in research and management of projects within the marine park (Figure 8.4).

[2] While the Authority retains the legal powers to grant or refuse hunting permits to Aboriginal and Torres Strait Islander people, the formation of Councils of Elders has established a *de facto* devolution of decision-making to local Aboriginal people with respect to the management of local dugong and turtle hunting.

Figure 8.4 Community ranger and government fishery officer.

8.9 IMPLICATIONS FOR OTHER MARINE PROTECTED AREAS

The GBRMP is widely regarded as a good model of large-scale marine ecosystem management, accommodating many human uses and providing opportunities for public input into management, but with regard to its accommodation of indigenous peoples' interests the GBRMP does not yet provide a model for others to follow. Nevertheless, there are valuable lessons to be learnt from the experiences of the Marine Park Authority in responding to the implications of managing a marine ecosystem which is part of living indigenous maritime cultures. If this or similar marine protected areas with significant indigenous peoples' interests were to be established now, it is reasonable to expect that the following measures would be addressed:

- Early and ongoing consultation with maritime indigenous peoples as to the nature and extent of their cultural, social and economic interests in the marine environment.
- Early and ongoing negotiations with appropriate indigenous peoples with respect to their meaningful involvement in planning and management of the marine protected area.

- Recognition of indigenous peoples' interest in all enabling legislation.
- Establishment of special management arrangements for the substantive involvement of indigenous peoples in the management of their customary domains and resources within the marine protected area.
- Recognition that the natural marine environment has intrinsic cultural values to indigenous peoples which must be accommodated in planning, management and interpretation of the marine protected area.
- Appointment and resourcing of appropriate specialist staff to facilitate ongoing liaison with indigenous communities, and to assist in the implementation of special management arrangements involving indigenous peoples.

For these measures to be implemented as a matter of course in marine protected areas in Australia, and other places where indigenous maritime cultures survive, considerable progress must be made in increasing awareness of the relationship between indigenous peoples and the maritime environment among the general community and among governments. Organizations such as the Marine Park Authority can benefit from this process and also have much to contribute to it.

If measures outlined in the 25-year strategic plan for the Great Barrier Reef World Heritage Area are promoted, the Marine Park Authority and the Queensland Department of Environment and Heritage have the opportunity of providing the sort of leadership in indigenous people's involvement in marine protected area management that they have already established in the management of the 'natural' marine environment.

It is worth noting, however, that significant Aboriginal involvement in the management of Australian terrestrial national parks only developed once formal recognition of Aboriginal ownership of the land was made. The Mabo high court decision has initiated legal and political processes that may see similar recognition of traditional ownership of marine estates, which in turn could be expected to lead to enhanced involvement of Australia's indigenous peoples in the management of marine protected areas.

REFERENCES

Baldwin, C. (1984) *Management of Dugong: an endangered marine food species of traditional significance. The need for user involvement in an integrated program of research, management and education*, GBRMPA, Townsville.

Bani, E. (1988) Badu and Thursday Islands, in *Workshop on Traditional Knowledge and the Marine Environment in Northern Australia* (eds F. Gray and L. Zann), Workshop Series No. 8, pp. 46–47, GBRMPA, Townsville.

Beaton, J.M. (1985) Evidence for a coastal occupation time-lag at Princess Charlotte Bay (North Queensland) and implications for coastal colonisation and population growth theories for Aboriginal Australia. *Archaeology in Oceania*, Vol. 20, No. 1, pp. 1–20.

Beckett, J. (1987) *Torres Strait Islanders: Custom and Colonialism*, Cambridge University Press, Sydney.

Bergin, A. (1991) Aboriginal sea claims in the Northern Territory of Australia. *Ocean and Shoreline Management*, **15**(3), 171–204.

Campbell, J. (1988) Role of Fishing in Aboriginal Society Before European Arrival in Australia, in *Workshop on Traditional Knowledge of the Marine Environment in Northern Australia*, (eds F. Gray and L. Zann), Workshop Series No. 8, pp. 1–3, GBRMPA, Townsville.

Chase, A.K. (1980) *Which Way Now?: Tradition, continuity and change in a north Queensland Aboriginal community*. PhD thesis, University of Queensland, Brisbane.

Chase, A.K. (1981) Dugongs and Australian Indigenous Cultural Systems: some introductory remarks, in *The Dugong*, (ed. H. Marsh), Proceedings of a workshop at James Cook University, Townsville.

Chase, A.K. and Sutton, P. (1981) Hunters and Gatherers in a Rich Environment: Aboriginal coastal exploitation in Cape York Peninsula, in *Biogeography and Ecology of Australia*, (eds A. Keast, R.L. Crocker and C.S. Cristian), Junk and Co., The Hague.

Cordell, J. (1989) Introduction: Sea Tenure, in *A Sea of Small Boats*, (ed. J. Cordell), Cultural Survival Inc., Cambridge, Mass., pp. 1–3.

Cordell, J. (1991) *Managing Sea Country – Report on Indigenous Fishing*, Ecological Sustainable Development Fisheries Working Group, Australian Fisheries Service, Canberra.

GBRMPA (1981) *Nomination of the Great Barrier Reef by the Commonwealth of Australia for inclusion in the World Heritage list*, Great Barrier Reef Marine Park Authority, Townsville.

GBRMPA (1994) *Keeping it Great: A 25 Year Strategic Plan for the Great Barrier Reef World Heritage Area 1994–2019*, Great Barrier Reef Marine Park Authority, Townsville.

Hale, H.M. and Tindale, N.B. (1933 and 1944) Aborigines of Princess Charlotte Bay, North Queensland. *Records of the South Australian Museum*, Vol. 5, No. 1, pp. 63–116; Vol. 5(2), pp. 117–172.

Hopevale Community Council (1990) Submission – Great Barrier Reef Marine Park Authority Cairns Section Zoning Plan Review.

IUCN, UNEP and WWF (1991) *Caring for the Earth – A Strategy for Sustainable Living*, IUCN, Gland, Switzerland.

Johannes, R.E. and MacFarlane, J.W. (1991) *Traditional fishing in the Torres Strait Islands*, CSIRO Division of Fisheries, Hobart.

Law Reform Commission (1986) The Recognition of Aboriginal Customary Laws, Part VII: the recognition of traditional hunting, fishing and gathering rights. *The Law Reform Commission Report*, No. 31, Vol. 2.

Lawrence, D. and Cansfield-Smith, T. (1991) *Sustainable Development for Traditional Inhabitants of the Torres Strait region – Proceedings of the Torres Strait Baseline Study*, GBRMPA, Townsville, pp. 283–294.

Mabo, E. (1988) Murray Island, in *Workshop on Traditional Knowledge of the*

Marine Environment in Northern Australia, (eds F. Gray and L. Zann), Workshop Series No. 8, GBRMPA, Townsville.

Marsh, H., Smith, A. and Kelly, G. (1984) *Dugong hunting by members of the Hopevale Community and the initial reaction to the GBRMP and the dugong permit system*, Great Barrier Reef Marine Park Authority Report, Townsville.

Nietschmann, B. (1989) Traditional Sea Territories, Resources and Rights in Torres Strait, in *A Sea of Small Boats*, (ed. J. Cordell), Cultural Survival Inc, Cambridge, Mass.

Pearson, N. (1988) *The myth of 'traditional user involvement in integrated programs of research, management and education' of Aboriginal sea: the need for Aboriginal control over usage of marine resources by Aboriginal communities.* Paper presented to the Innovative Planning and Management Workshop, July 1988, GBRMPA, Townsville.

Rigsby, B. and Williams, N. (1991) Re-establishing a Home on Eastern Cape York Peninsula. *Cultural Survival Quarterly*, Vol. 15, No. 2, pp. 11–15.

Smith, A. (1987) *Usage of Marine Resources by Aboriginal Communities on the East Coast of Cape York Peninsula*, Report to the Great Barrier Reef Marine Park Authority, Sir James Fisher Centre for Tropical Marine Studies, James Cook University, Townsville.

Smith, A. (1989) *Usage of Marine Resources by Aboriginal People on the East Coast of Cape York Peninsula*, GBRMPA, Townsville.

Smyth, D.M. (1989) *Aboriginal Liaison – assistance for Aboriginal communities as part of the review of the Cairns section zoning plan*, Consultancy Report, GBRMPA, Townsville.

Smyth, D.M. (1990) *Aboriginal Maritime Culture in the Cairns Section of the Great Barrier Reef Marine Park*, Consultancy Report, GBRMPA, Townsville.

Smyth, D.M (1992) *Aboriginal Maritime Culture in the Far Northern Section of the Great Barrier Reef Marine Park*, Consultancy Report, GBRMPA, Townsville.

Smyth, D.M. (1993) *A Voice in All Places: Aboriginal and Torres Strait Islander Interests in Australia's Coastal Zone*, Resource Assessment Commission, Canberra, Australia.

Sutherland, J. (1992) Rising Sea Claims on the Queensland East Coast. *Aboriginal Law Bulletin*, Vol. 2, No. 56, pp. 17–19.

Thompson, D.F. (1934) The Dugong Hunters of Cape York. *Journal of the Royal Anthropological Institute*, Vol. 64, pp. 237–264.

Wallace, J. and 12 Community Rangers (1992) Aboriginal Involvement in national Parks: Aboriginal Rangers' Perspectives, in *Aboriginal Involvement in Parks and Protected Areas*, (eds J. Birkhead, T. de Lacey and L. Smith), Australian Institute of Aboriginal Studies, Canberra, pp. 29–38.

Ziegelbauer, W. (1991) *Aboriginal Interests in the Great Barrier Reef*, Report to the GBRMP Consultative Committee, Townsville.

9

Marine education and interpretation

Stephanie Kaza

Humpback whales breaching in the sea, gooey slicks from massive oil spills, strange machines for underwater marine research, unsightly litter and plastics on the beaches – through environmental education efforts, these are now well-known images in some areas. But for many people and cultures around the world, the oceans are still mostly 'out of sight and out of mind'. Yet we know that the lands and the waters are both critical parts of the ecological functioning of this planet. Ocean weather, currents and food webs greatly influence land habitats, organisms and geophysical processes.

As land resources dwindle and coastal populations swell, there is a great need for awareness of human actions and their impacts on the oceans. This is not only desirable and useful; it is a moral and economic mandate, especially for industrial nations who are responsible for much of the worldwide marine pollution, global overfishing and other forms of waste disposal and exploitation. Marine education and interpretation are essential if we are to understand and sustain the environments in which we live.

Interpreting marine protected areas is easier said than done. Because conservation and management of marine ecosystems is still relatively new, our collected experience in field-based marine education is relatively small. Constraints and challenges abound in dealing with issues of safety, access, conservation and communication. Because of the fluid nature of the site, the lack of stable substrata and changeable conditions of weather or sea state, techniques developed for land-based situations are often not applicable or not appropriate. To add to the challenge, marine protected areas are inevitably underbudgeted, leaving few funds for education staff or resources. Nonetheless, with creative planning, ingenuity and dedication, marine interpretation can be extremely effective in developing public support and understanding of the marine environment.

9.1 DESIGNING AN EDUCATION PLAN

Public education efforts in a marine protected area can easily become scattered and fragmented without some sense of overall strategy and design.

Marine Protected Areas: principles and techniques for management. Edited by Susan Gubbay.
Published in 1995 by Chapman & Hall, London. ISBN 0 412 59450 1.

There is a tendency to focus on individual organisms or specific areas of the reserve without developing an overall message or sense of the whole system. An education and interpretation plan can provide direction and priorities for action, with a conscious intention to build public understanding and community involvement. The first task of planning is focused on surveying the territory and asking these questions: what is there to interpret, what tools and staff are available for interpretation and what management problems exist that could benefit from public education?

9.1.1 The site

At the heart of a protected area are its physical, biological and cultural resources. The most fundamental objective of an education programme is to communicate the importance of key aspects of the site and why they merit protection and attention. Interpretive programmes then include discussions of vulnerable habitats, common and endangered species, ecological relationships and special physical features. At an international training seminar sponsored by the United States National Marine Sanctuary Program, some of the topics of concern were turtle nesting areas, endangered fish, coastal headlands and coral reef zones (Kaza, 1988). Participants found single species to be easier subjects for interpretation than whole ecosystems, and ecological relationships were the most challenging teaching subjects. A thorough assessment of resources provides a catalogue of possibilities for interpretation from which to design an appropriate education programme.

Cultural aspects of the site are equally important: historic uses or artefacts, local economic uses, scientific and recreational value and spiritual importance to indigenous or transplanted peoples. Cultural relationships discussed at the seminar included fishing, tourism and territoriality, with some class and racial conflicts over priority values. The site survey is thus important in laying the groundwork for understanding the value of an area in terms of both biological diversity and cultural participation.

9.1.2 Staff and resources

The second element of a planning survey is a realistic assessment of staff and materials available for education. In almost every situation we considered, resources ran far short of what was desired or needed. Some marine protected areas had exhibits, posters or brochures but few had school activities, slide programmes, species lists, trail guides or public relations materials. Visitor centres are small, if present at all, except for a couple of large marine sanctuaries in the United States and Australia. Most areas have no education staff or volunteer programme, limiting the possibilities for personal contact with the public.

However, quite often this lack of resources can be mitigated by drawing on other community or government resources. We looked closely at potential community groups which could work cooperatively with the protected area, and at key local leaders who might be influential in creating public support. Inevitably there is some kind of community response to the establishment of a new marine reserve. The challenge for administrative staff is to cultivate positive involvement from the start, before local groups and individuals have taken a stand against the reserve. In some cases, government agencies with overlapping jurisdictions, such as the state Departments of Fish and Game in the United States, or adjacent national parks, can provide operations support and staff consultation. Where joint efforts enhance a community group or government agency as well as the marine protection areas, these education projects may be implemented more quickly.

9.1.3 Management concerns

Often an area is set aside for protection specifically to deal with harmful activities. Are local citizens overfishing in critical habitat areas? Do poor agricultural practices upstream pollute river run-off into a protected estuary? Are shells and coral samples disappearing to tourists at an alarming rate? Do local agencies disagree on protection goals for the area? These kinds of management problems are often the first targets for education efforts. Simple cause and effect explanations on public flyers can reduce litter and insanitary practices. Posted signs can provide guidelines limiting over-collection of endangered species.

Many management problems or conflicts fall into the category of 'growing pains' of a new sanctuary or reserve. Just getting started has its own inherent hurdles or resistance. Sometimes a site's history, geography or politics present unusual challenges that simply go with the territory. Each area requires a creative and realistic appraisal of what is possible. But once an education programme is in place, it is relatively easy to build on success and accumulate useful materials and ideas.

9.2 KEYS TO A SUCCESSFUL STRATEGY

Given the resources and problems in a specific marine protected area, the next step in planning is to articulate education objectives. These are likely to include some management objectives that can be accomplished through education techniques. The site resources together with the objectives may suggest educational themes such as ecological interaction, adaptation and evolution, patterns of change and distribution, or habitat conservation. Proposed education programmes should be aimed at specific audiences and characterized by clearly identified messages. The educator's challenge is to choose the most effective educational technique for communicating the message to the targeted audience.

9.2.1 Education objectives

Education objectives provide general guidelines for programme content and orientation. In the US Gulf of the Farallones National Marine Sanctuary, the programme's mission is:

- to foster ecological awareness and promote appreciation for biological and cultural complexity; and
- to develop patterns of behaviour for living responsibly and sustainably with the marine environment.

More detailed objectives suggest levels of understanding and supporting concepts that describe the web of relationships in the areas (Kaza, 1985). In this new reserve, primary education objectives were to maximize public contact, to build public support and programme recognition and to promote regional identification with the marine environment.

An interpretive programme can also be based entirely on critical management objectives related to the health of the site. These might be:

- To maintain and enhance biological diversity.
- To preserve the local marine cultural heritage.
- To develop an economic base in marine tourism.
- To maintain ecological processes and life support systems.
- To promote sustainable levels of resource use.
- To promote government efforts at marine environment protection.

The interpretive message should reflect these objectives in a simple and direct way. Three or four key ideas are plenty in establishing the educational thrust of the programme. They may be as straightforward as: 'Let's take care of where we live,' or, 'Give a hoot, don't pollute.'

9.2.2 Audience

Identifying target audiences is the key to maximizing the effectiveness of existing staff and material resources. While traditional education programmes are often aimed at school children, in its first years a marine protected area may need to orient its programmes towards local leaders, government officials or influential community members. If specific plants or animals are being overcollected or harmed by local residents, education efforts should be aimed directly at those causing the damage. For example, in the Galapagos, the rare black coral so popular with tourists is now monitored carefully to restrict sales and rates of collection by local islanders. In California, whale-watching guidelines for boat operators printed by the marine sanctuaries help reduce grey whale harassment, and a brochure on safe dumping suggests ways to reduce garbage and plastic at sea.

If the general public is the target audience, the education goals may be simply to promote the marine environment as worthy of attention and conservation. As well as by personal contact, this may be most effectively accomplished through mass media such as radio and television interviews or newspaper articles. The point in each case is to find the appropriate educational vehicle for conveying the message to the selected audience. If locals are overcollecting tropical fish, it may not necessarily help to explain the issue to visiting tourists. If a marine area is in need of government support, it may do little good to carry the message to school children. The effectiveness of an education programme depends on understanding the nature of human relations and how people work with information and values.

9.3 EDUCATION TECHNIQUES

In choosing education techniques for a marine protected area, the keys to success are adaptability, diversity and creativity. A technique that works well at one site may be adapted to some others but may be quite inappropriate for another site. Preferred modes of learning will vary from country to country. Education technologies such as slide projections and video films may be standard tools at some sites, while others will be lucky to have a typewriter. There is no single educational model that suits all situations. The important point is that the chosen methods work in the given situation and are effective with the audience at hand. Each element of an education programme must be constantly evaluated to see if it is successful in accomplishing management and education objectives.

The range of interpretive tools is broad and varied, each with its own advantages and disadvantages in relation to a particular site or culture. A number of education publications discuss these techniques in detail with specific examples and guidelines for projects and staff development (for example, see Berkmuller, 1981; Sharpe, 1982; Wood and Wood, 1986). The following sections provide an overview of six types of educational activity, with reference to some of the more successful examples in several marine protected areas.

9.3.1 Personal contact

The person-to-person approach is very effective in conveying information, conservation values and agency presence to visitors and local community members. The experience of a personal exchange can make an impression that lasts beyond a single visit to a protected area. Through direct contact, people can become inspired to learn and care about the marine environment. Personal contact in the form of tours or prepared talks is common in developed countries, especially in state and national parks. These formats can translate to marine parks, especially in urban areas or sites with high visitation. Professional staff trained in communication techniques and marine ecology can serve as a link between the site and the visitor, sharing information and concepts that will broaden public understanding.

Most marine parks, however, do not have funds to support full-time educational staff. In the United States, the largest marine sanctuaries have only three or four staff members to cover resource protection, enforcement, administration, planning and also education. Relative to other techniques, personal contact is an expensive budget item and requires people with developed skills. However, volunteers can be of great assistance and generate community participation and interest in a protected area. For example, Parks Canada plans to involve local native Indians to interpret tribal lands and waters at its park at South Moresby Island, British Columbia. At Año Nuevo State Beach an elephant-seal breeding reserve in central California, more than 100 volunteer **docents** (helpers) and college students host 50 000 visitors each winter in an average of 25 tours each day, a scale of interpretation that could never be managed by park staff alone.

Even if staff resources are limited, they can be invested in training others who work on behalf of the marine environment. This may take the form of teacher workshops, agency-wide seminars, or interpretive training for co-operating non-profit groups. In the San Francisco area, marine sanctuary staff have provided educational workshops for whale-watching naturalists on local conservation issues. Park staff at Santa Cruz, California, have offered training for student volunteers and community docents to interpret the local rocky intertidal at Natural Bridges. The enthusiasm and commitment of staff or volunteers naturally serve to spark others' learning and interest in the marine environment.

9.3.2 Visitor facilities

Exhibits, trails, wayside signs and visitor centres provide a focal point for a site's identity. Where staff are not available, physically constructed facilities offer interpretive contact and an identity and image for a marine protected area. For example, the visitor centre at the Channel Islands National Park and Marine Sanctuary is a large, stunning building on the coast of Santa Barbara, commanding a sweeping view out to the ocean and the islands. Its presence alone makes a statement about the importance and value of the islands and surrounding marine waters. In contrast, the modest visitor centre at Tortugera National Park in Costa Rica is a homely shelter out of the rain, where people can sit down for a cup of coffee with local boatmen and look at rainforest posters.

In a few places underwater trails have been developed for scuba-divers and snorkellers. At Point Lobos, California, the trails help to direct traffic for the high number of visitors to this popular dive spot while also illustrating the natural history of subtidal invertebrates and marine plants. Some marine protected sites, such as the Monterey Bay Sanctuary in the United States, lie adjacent to protected land parks so shoreline trails can provide ocean-oriented interpretation for recreational hikers.

Wayside signs and exhibits can be effective aids for the self-guided visitor but they are subject to weathering and vandalism. Salt air, wave action, intense sun and high winds all take their toll on constructed signs. Illustrations must be kept simple; photographs weather quickly. Only the toughest of paints and construction materials will last for more than 5 years. If the local community does not support the area's protected status, they may remove or damage signs to undercut the agency's presence. To counteract this possibility, it can be helpful in some areas to involve the members of the community in planning, designing and constructing visitor facilities. In this way, the centre can serve the local community's needs as well as those of tourists.

9.3.3 Printed materials

Brochures, posters, pamphlets and books can provide important information about marine conservation and ecology and also help to establish an image for a marine protection area. Brochures are introductory pieces that give a visitor or agency official some sense of the unique features of the area and why protection is important. Complex brochures may also offer useful boating information, maps and natural history information about common species. The task of producing a brochure or pamphlet, no matter what the scale, helps an agency to define its mission for the public in understandable terms. More detailed guidebooks are a welcome addition to the educational library of both tourists and local community members concerned with the marine environment. Even environmental comic books can be helpful, as in the Caribbean, where they are a key teaching tool for school children. Books that tie together the many ecological and cultural aspects of an area, such as those written about Point Reyes (Evans, 1988) and Año Nuevo in California (Le Boeuf and Kaza, 1982), are particularly useful in presenting a systems approach to protection.

Glossy or inexpensively produced posters can promote community support for the marine environment by presenting beautiful and inspiring images of local fish, birds or other marine scenes. In Malaysia, posters of lovely coral reef fish caught the much needed attention of government officials in time to achieve area protection. In California, posters of seabirds and marine mammals are well-used references in local elementary classrooms. Poster images can often be reproduced on T-shirts, so that the park's message is distributed in a diversity of ways. In Fagatele Bay in American Samoa, the National Marine Sanctuary distributed a beautiful tide calendar, illustrated with children's drawings from the local community.

In all printed materials, choice of language and graphics can make an enormous difference to whether a piece is used or ignored. If a reader has to struggle with small print, overly detailed drawings or inaccurate maps, the educational message might get lost. If a brochure is produced in English but not the local language, it might be seen as non-inclusive or, at the worst, offensive to indigenous peoples. If the only illustrations are of certain races or subcultures, it might imply that they are the only users or abusers of the marine area. Each graphic element in a printed piece carries weight and cultural as well as visual value, creating a concrete impression that can serve the long-term success of the marine protected area.

The advantage of printed pieces is that they can be taken away from the site and serve as a representation of the area for decision-makers and potential visitors. Large numbers of brochures and posters can be printed relatively inexpensively, depending on paper and printer availability. A park's message can be stated clearly and directly, in unmistakable terms, with essential information provided. However, it is important to note that not all users or visitors to marine protected areas are literate and it would be inappropriate to rely on printed materials in some situations.

9.3.4 Mass media

Film, television, radio and newspapers have the advantage of being able to reach large numbers of people. Coverage of environmental issues is increasingly common and many natural history stories are of high value in the entertainment ratings. Because of their unusual adaptations to life at sea, marine animals make good features; whales have been a consistently popular topic for the last 10 years along the whale-watching coasts of the United States. Visually effective material, such as the footage of humpback whales on Silver Banks in the Dominican Republic, can help to create widespread public and government interest in marine area protection. Promotional clips also generate interest and financial support from wealthy developed countries concerned about conservation and willing to fund protection in other hemispheres. World Wide Fund for Nature and Conservation International have often relied on film footage for documentation of candidate areas for protection.

Choice of mass media technologies generally reflects the size of the budget for the protected area and the scale of urban development in the nearby region. With a little initiative, the effective manager can generate extensive publicity coverage by local television and newspaper reporters. Personal contacts with media people are perhaps some of the most important relationships to cultivate in the early stages of a park's existence. In-house film and video production, however, requires expensive equipment and processing and most protected areas will not be able to afford such single-item expenses on their own.

9.3.5 Working with schools

Contact with teachers and students at all levels of education reaches right into the heart of the community and serves as an avenue for contact with parents and other adults. In some areas, children may be the most receptive to information about marine ecology and conservation and can educate other generations in a multiplier effect. Where enthusiasm runs high, park staff or volunteers will want to work in schools because of the community interest it generates. While production of school material may not be a top priority for new marine protected areas, this is not necessarily a significant barrier to effective education. Children can generate their own learning materials from reports, lab drawings and field observations.

There are a number of effective marine education curricula in the United States that can be used as models for school-based education in other regions (Project Aquatic Wild, For Sea, Alaska Sea Week, among others). Most of these are based on an experiential rather than a didactic approach to learning, engaging students directly with the site and the plants and animals that live there. They promote an environmental ethic that reflects a sense of belonging rather than separation from the natural world. Marine education easily cuts across all subject areas and can be adapted to suit a teacher's interests and curriculum needs, no matter what the culture. Tide tables become mathematics lessons, algae pressing is an art project and clam dissection teaches science principles. Though children do not sit in the seats of power, their excited smiles can sometimes have a powerful influence on key decision-makers responsible for the fate of marine protected areas.

9.3.6 Community outreach

Special events, extension programmes and liaisons with local leaders of non-governmental organizations may have the most direct influence in building popular support for a marine protected area or changing local traditions that harm the marine environment. Community outreach can draw on the strength of already established groups, working with the network of human relations that influences local culture and behaviour patterns. In the UK, for example, organized clubs of scuba-divers learned about local marine life and gained some awareness of conservation issues by participating in a nationwide sea urchin survey. In some areas community-based programmes can make a major impact on the health of a marine system. In the state of Oregon, more than 5000 citizens joined in a major beach clean-up campaign to get rid of unsightly debris; the programme is now a model for other states' efforts.

Community outreach with specific target audiences can increase local awareness and generate greater involvement in marine conservation. Fishing villages in the Philippines benefited from community marine ecology classes oriented towards local use of marine resources. Marine advisers then facilitated the development of local marine management groups that took primary responsibility for self-regulation practices. Staff at the Gulf of the Farallones National Marine Sanctuary in California engaged community leaders and local agencies in developing a plan for needed research. Community outreach requires sensitive diplomacy and consistent effort, but it is often the most important arena for interpretive efforts.

The best source of ideas and strategies for marine education is other programmes and staff. A good interpretation programme takes the best features of existing programmes and adapts them to the situation at hand. In the following sections three cases studies of highly successful marine programmes in the San Francisco Bay area are reviewed. Each has strengths to offer and educational elements that can be modified for other areas. Each is an example of effective interpretive planning in a different educational setting – a protected site, the classroom, and a public museum.

9.4 MODELS FOR MARINE EDUCATION

9.4.1 Gulf of the Farallones National Marine Sanctuary

Just offshore from one of the world's major metropolitan areas – the San Francisco Bay area – lies the Gulf of the Farallones, a biologically and culturally rich marine protected area. The Gulf is part of the California Current system of the Eastern Pacific, influenced by the strong seasonal upwelling of cold ocean bottom water. This coastal area receives water from over one third of the state of California, flowing down the northern Sierra Nevada across the Sacramento Delta through San Pablo and San Francisco Bays. National marine sanctuary status, awarded in 1981, extends protection 12 miles (19.3 km) around the Farallon Islands and 6 miles (9.7 km) along the coast, encompassing 948 square nautical miles of ocean waters.

From the northern boundary at Bodega Head to Rocky Point south of Stinson Beach, the Sanctuary includes an unusual diversity of marine and estuarine environments – protected bays, rocky intertidal zones, sandy beaches, open pelagic water and the deep benthos of the continental slope. These waters support the largest concentration of breeding sea-birds in the continental United States. Many of the 22 species of marine mammals occur in large concentrations and are dependent on the productive and undisturbed waters of the sanctuary and adjacent coastal areas for breeding, pupping, hauling-out and feeding. The area is also significant for its many species of marine fish and invertebrates, reflecting the diversity of habitats.

The Gulf of the Farallones is a superb showcase for the physical and biological processes of the Pacific marine environment. Over 58 coastal points in Sonoma, Marin, San Francisco and San Mateo counties provide a direct view or marina access to this sanctuary. To the north, the protected area is contiguous with Cordell Bank National Marine Sanctuary; to the south, it is very close to the site of the proposed Monterey Bay National Marine Sanctuary. This unusual combination of protected areas provides an outstanding platform for marine education and interpretation along the central California coast.

The need for planning

When the site was designated in 1981, the National Marine Sanctuary Program and the Gulf of the Farallones were little-known entities in the minds of the Californian general public. The site was enormous and the challenges of rough water, dispersed access sites and minimal institutional presence and staff made education and interpretation difficult. It was important to determine what was possible amd appropriate and what would be most helpful in promoting the sanctuary and appreciation for the marine environment. Given the limits and nature of this particular protected area, education activities were strategically chosen for maximum effectiveness, within a carefully designed plan of goals and objectives.

Under contract to the sanctuary, this author developed a plan that could serve the protected site for the next 10 years and be a model for other site educational planning efforts (Kaza, 1985). Because of the lack of education staff (only two management positions existed), the obvious direction was to work cooperatively with other government agencies and marine education and conservation groups. Research for the plan was based on three surveys of: local marine organizations; potential and current user groups for the sanctuary; and interpretive sites.

The wider metropolitan region of San Francisco Bay hosts a large number of private, professional, commercial and non-profit groups dedicated to marine activities and education. The study evaluated groups for potential cooperative roles with the sanctuary according to four criteria:

1. Degree of overlap of the organizations's purpose with the sanctuary's purpose, especially regarding management and education objectives.
2. Degree of public influence the organization held according to its size and public programmes or political strength.
3. Capability of the organization to carry out cooperative programmes.
4. Degree of commitment to the National Marine Sanctuary Program and the Gulf of the Farallones.

From the original list of 23 organizations in the survey, seven were selected as key resources for joint programmes; the remainder were relatively less likely to be involved in sanctuary projects.

The survey of user groups drew together information on those people seeking an active experience with the marine and estuarine environments of the sanctuary. Activities clustered under:

- educational recreation (bird-watching, whale-watching, tidepooling, art/photography, hiking);
- pleasure recreation (beach use, boating and sailing, paddling, sport-fishing, clamming and abalone diving, waterfowl hunting);
- commercial uses (fishing, charter boat operations, aquaculture, shipping);
- education (school field trips, adult education programmes, research).

The survey identified groups with the greatest interest in marine conservation and possible education methods or opportunities for each type of user group. Because of the diversity of visitors to the sanctuary, the plan recommended targeting specific groups to provide appropriate educational materials.

With 58 potential sites for sanctuary interpretation, it was necessary to rank these sites for maximum effectiveness and overall educational value to the sanctuary. Using local coastal guides and field visits to the sites, we ranked the sites according to:

- accessibility – the degree of real or perceived difficulty in getting to the interpretive site (as judged by the quality of roads, availability of support services such as food, fuel and lodging, and the remoteness of the site);
- biological and/or cultural richness – based on diversity of habitats or cultural landmarks, the size of the feature, and the complexity or uniqueness of key interpretive attractions;
- current use level – the current popularity of the site as an indication of the established flow of visitors.

Key sites with high ranking were popular parklands adjoining the sanctuary and visitor attractions outside the sanctuary with strong marine interpretive programmes (Figure 9.1).

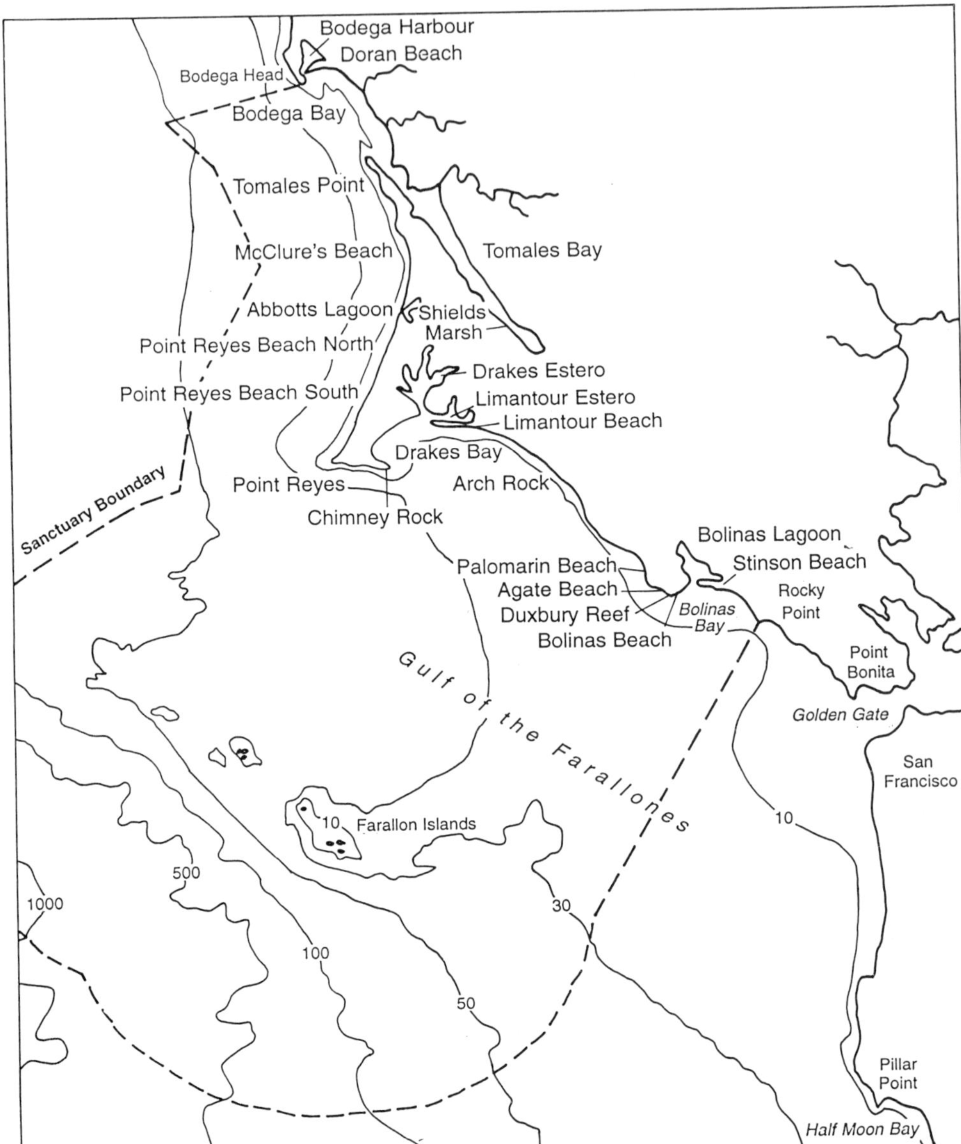

Figure 9.1 Key and secondary sites for interpretation in Point Reyes–Farallon Islands National Marine Sanctuary (from Kaza, 1985).

Educational objectives and accomplishments

Given this extensive range of interpretive sites, audiences and cooperating organizations, educational objectives were essential to set a direction and serve as criteria for selecting and funding specific projects. We proposed three overall objectives:

- Increase public awareness of the sanctuary, current marine protection issues and the National Marine Sanctuary Program.
- Offer opportunities for first-hand experience, appreciation of and involvement with the sanctuary.
- Develop a strong network of local support through cooperative efforts and public information.

Over the first 3 years, the strategy would be to focus on broad-level public outreach and orientation, building sanctuary identity and recognition of the ecological significance of the sanctuary. The early years were important for maximizing public contact, ensuring a sustainable future for the sanctuary and developing a positive and professional image for this federal programme. The next 3 years would build on this foundation and focus on interpretive activities and materials that address specific needs. The next 6–10 years were designated for implementation of large-scale projects and evaluation of the programme as a whole to determine directions for the future.

The educational and interpretation plan was completed in 1985 and the programme was well into the second phase of its objectives by 1990. What had been accomplished? Did the plan provide useful and appropriate guidelines for choosing projects? In fact, the sanctuary was highly successful in collaborating with local organizations on a marine mammal poster, a whale-watching checklist, a marine science slide programme for teachers, public symposia, school curriculum and site visits, and whale-watching and boating guidelines. Staff produced the first glossy colour map brochure of the National Marine Sanctuary system, establishing new standards for the programme's public image. They developed basic interpretive tools useful in a diversity of settings. They worked closely with government agencies, marine science education programmes and conservation organizations to establish a presence in the Bay area. After only 5 years of active education programmes, the sanctuary had become a leading agency in the local area for promoting awareness of the marine environment of the central California coast.

Strengths of the programme

As models for other MPAs, the Gulf of the Farallones NMS offers a method of planning and implementing educational objectives, based on availability of interpretive sites, cooperating agencies and sanctuary staff. The thorough research of marine-related organizations in preparing the plan introduced the marine sanctuary and laid the groundwork for future interactions with these groups. The user survey provided an overview of the key members of the general public involved in marine-related activities, revealing dominant interests in education, recreation and commercial activities. The site survey gathered information helpful in focusing interpretive efforts by location and accessibility. By surveying the broad view of the sanctuary from the beginning, education efforts could be focused on immediate objectives with maximum impact.

The education plan served as a compendium and guideline for interpretive ideas, by no means exhaustive and certainly with room for creativity. The list of accomplished projects to date reflects not only the plan but also the expertise and interests of staff, the proposals of various cooperating agencies and contingencies of real life, such as the massive oil spill of the Puerto Rican off the Farallon Islands. Planning helped to provide continuity for the programme through changes in staff and facility location, as the sanctuary was becoming established. The plan itself was an educational tool for government agencies and organizations involved with the sanctuary. Information technology (IT) helped to provide a focus, a place to begin and a way to be effective with few resources.

By working collaboratively with other organizations, the sanctuary encouraged joint responsibility and participation in caretaking and interpreting the marine environment. By paying attention to a diversity of user groups, the sanctuary established a presence in many areas of marine activity. By working in a broad range of media – from printed materials to personal contact, and from news media to wayside signs – the sanctuary reached out to people in many ways. While this sanctuary is less remote than some marine protected sites, and is very fortunate in having a number of capable cooperating organizations, the planning approach used here is still applicable elsewhere. For at the heart of it, the plan is based on a thorough and realistic assessment of the nature of the site and its human resources, looking at what is actually possible and most needed.

9.4.2 MARÉ

Walk into a MARÉ classroom and you will see fish paintings on the wall, a kelp forest hanging from the ceiling, a squid being dissected by children and teachers who never dreamed they could do such a thing. MARÉ (Marine Activities, Resources and Education) is a highly successful marine science education programme based in the San Francisco Bay area at the University of California, Berkeley's Lawrence Hall of Science. It was founded in 1991 and developed from the award-winning Project OCEAN programme (Strang, 1989).

MARÉ is based on hands-on, school-wide participation from teachers and students at the elementary age level. The approach is interdisciplinary, with activities that use the ocean as a central theme. Each grade focuses on a different marine habitat, developing a unit of lessons over several weeks, months or the entire school year. Experientially-oriented activities enhance language acquisition and confidence-building for students with special needs. The programme is non-traditional, in that it is based not on any classroom textbook for children but rather, on direct engagement with the material and processes of marine science.

Elements of the programme

The centrepiece of the programme is Oceans Week, an intensive week of activity in which the ocean becomes the focus of the entire school curriculum. Mathematics classes graph the tides, art classes build whales, science classes test water quality, English classes read sea stories and all students attend a marine science field trip. The whole school becomes an integrated laboratory for studying the ocean. Each class prepares for the week of 'immersion learning' by studying a marine habitat in depth (Figure 9.2). MARÉ staff work closely with teachers and administrators during the week – not by teaching classes themselves but by coaching resident teachers,

Figure 9.2 Class using microscope.

leading mini-workshops and coordinating guest presenters for the week. These include the Whale Bus with its collection of whale bones, the Banana Slug String Band with its marine science rap music and the Monterey Bay Aquaravan with its travelling puppets and marine animal costumes. Some schools become so excited about Oceans Week that the following year they elect to hold an Oceans Month.

By engaging the entire school in a single focus, the climate and culture of the school improves as teachers and students work together with enthusiasm. Evaluation reports have shown that attendance often increases during this time, and some schools report that racial incidents and vandalism actually decrease with the increase in student excitement. Oceans Weeks have been held in such diverse places as inner city black schools, wealthy suburban schools, rural farming communities, the desert environment of Nevada and schools with a diversity of Asian and Hispanic minorities. Because the programme is oriented towards active hands-on learning, students gain cooperative skills and are not limited because of lack of English language proficiency (Figure 9.3). The immersion week is the highlight of a year-round focus on marine science with an integrated curriculum that teaches basic principles of physical and biological science.

To prepare for Oceans Week and the year-round marine science curriculum, MARÉ offers all-faculty in-service workshops and a summer training

Figure 9.3 Japanese fish printing.

institute for delegate teachers (Figure 9.4). These sessions provide teachers with current marine science information on a comprehensive range of topics, from underwater exploration to endangered species. They introduce teachers to innovative teaching strategies through demonstrations of curriculum activities. During the 10-day teacher institute, teachers develop plans for their own classroom units as well as the school-wide Oceans Week. These representative teachers then return to their schools to assist with the faculty-wide training. Throughout the year, MARÉ staff offer enrichment sessions that introduce new activities or provide field experience for enthusiastic teachers.

The third key element of the programme is the curriculum – a set of grade level specific Marine Habitat Guides, filled with hands-on activities, projects, background information and references. The guides reflect a systems approach to science, using marine habitats as a focus to show relationships between earth, physical and biological sciences. Each 200-page guide contains activities from these content areas as well as 'Human Interaction with the Marine Environment', where the focus is pollution, conservation, environmental ethics and cultural use and value of the ocean. Activities from a diversity of marine science curricula have been field-tested and adapted for effectively addressing the systems relationships of each habitat.

Educational philosophy

The beauty and strength of this programme as a model for others is its adaptability for a variety of students, its commitment to improving teacher capability in science and its coherent ecological curriculum. In the United States, 80% of Americans will live along a coast by the end of the century, yet few schools provide instruction in marine science. Recent tests in California show that ocean science is among the lowest science scores in student achievement. The difficulty lies not only in the general lack of available or appropriate materials but also in a lack of substantive training for teachers. As with many areas of science, the pressing need is to change the foundation of science instruction from transmission of facts to the skills and processes of scientific inquiry.

From the start, MARÉ has been committed to engaging all students in its programmes, including those 'at risk', disadvantaged students and limited English-proficient students (which constitute up to 50% of many Californian elementary classrooms). To ensure this level of participation, lessons are designed around activities rather than deskwork and students frequently work in cooperative groups or with lab partners. Activities teach vocabulary through dissection or simulation games, giving all students equal opportunities for building science language capability. The whole school immersion

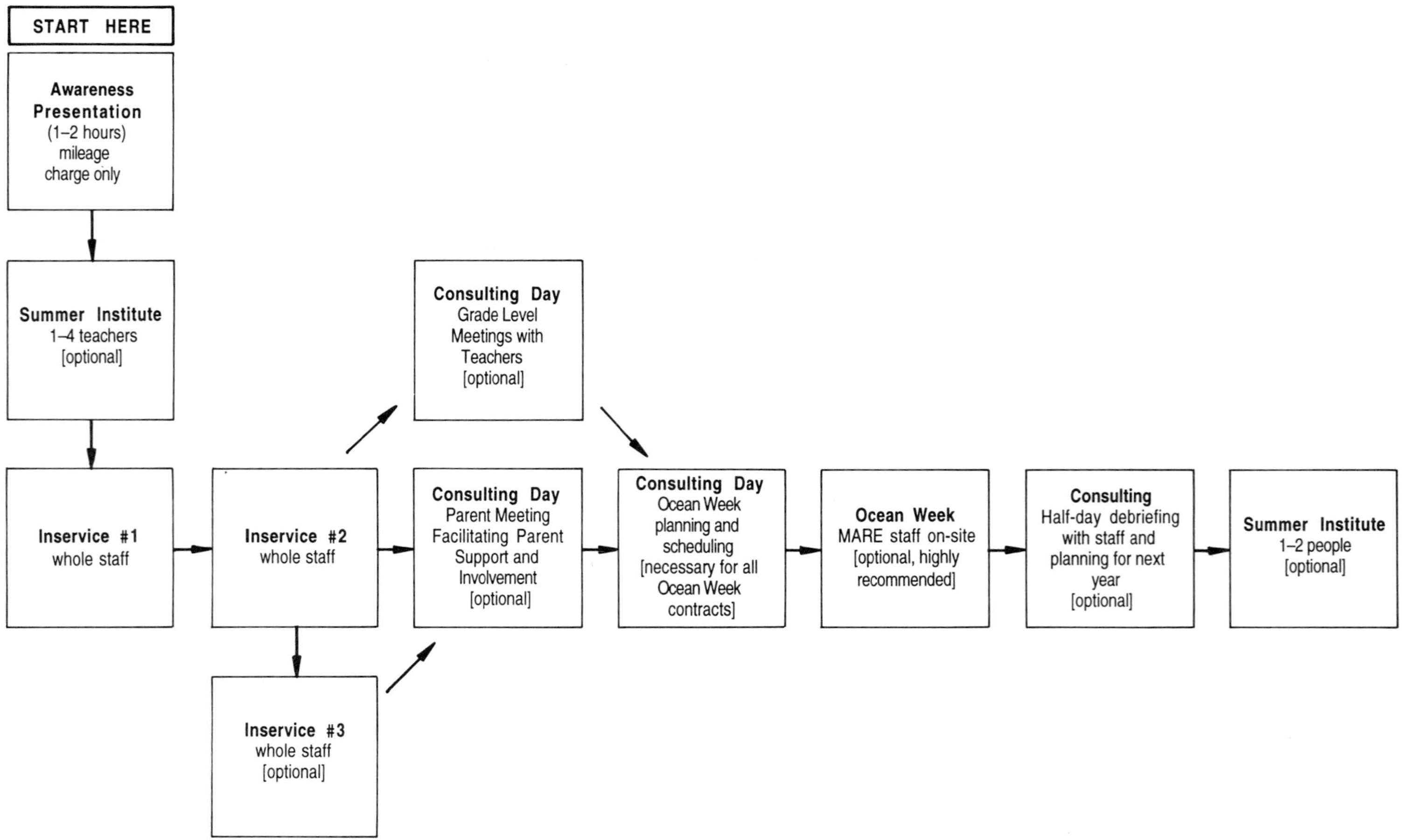

Figure 9.4 Typical first-year MARÉ school programme.

aspect of Oceans Week enthuses all students, even those with lower confidence, and maximum learning occurs as students take an active role in their own education.

This intensive programme and shift in curriculum emphasis requires a relatively extensive amount of teacher training. Teachers generally must first go through the same learning process expected of students before they feel confident about the subject matter. MARÉ addresses these needs by:

- enhancing instructional strategies with an emphasis on student engagement and responsibility;
- helping teachers to plan and improve their own science curricula;
- coordinating peer training to build professional support for change.

The training institutes provide practice sessions with specific activities and exposure to high quality instruction. Staff offer in-school and external support throughout the year to institute teachers and participating schools. To ensure a long-term commitment to the programme, MARÉ encourages development of a local support system within the school, partnerships between school districts and administrative involvement. Ideally, teachers can be guided through one to three years of training and support from MARÉ staff, and then be on their own.

The ecological framework of the curriculum is fairly unusual among marine science programmes. Core physical, biological and human interaction themes are carried from grade to grade and addressed across the habitats. Each activity presents a key focal concept which is integrated with other concepts through transitional statements and example learning sequences. Teachers can freely order their lessons according to personal or student interest, building a unit across the disciplines by linking key concepts. The point is to study the habitat from a variety of perspectives and understand it as a living system rather than a collection of disparate elements.

The marine habitats represent the most common ecosystems along the eastern Pacific; the curriculum is designed to be relevant to the local bioregion and to encourage a sense of identity with where students live. For schools located inland, west coast habitats are still the most relevant. The curriculum has also been adapted to the Gulf coast of Texas and east coast habitats. Habitats are correlated to grade level by complexity and teachability; Rocky Seashore (first grade, or ages 6–7), Sandy Beach (second grade), Wetlands (third grade), Kelp Forest (fourth grade), Open Ocean (fifth grade), Islands (sixth grade), Coral Reefs (seventh grade) and Polar Seas (eighth grade).

MARÉ offers an effective and coherent model for marine science curricula elsewhere because of its emphasis on experiential learning, local marine ecosystems and culture, and through teacher training. The immersion experience of Oceans Week can be implemented anywhere and serve as a powerful starting point for school-wide learning. Curricula based on similar principles can be designed for local marine regions. While MARÉ staff have been important in implementing the programme, in most places, the scale of success is due to enthusiastic teachers and students within individual schools who have a great time learning about the sea around them.

9.4.3 Monterey Bay Aquarium

When the Monterey Bay Aquarium opened its doors in October 1984 after 4 years of construction, it set a new standard for public aquaria in the United States. Engineers were astonished at the scale of the exhibits, two of which held over 300 000 gallons of sea water; visitors were thrilled with the diversity of displays and interactive galleries; and the local community recognized an outstanding cultural institution, now drawing people from all over the world. Thousands of hours of planning and preparation went into the design of the exhibits, which focus entirely on the rich marine life of the Monterey Bay, California (Gore, 1990).

More than 6500 specimens representing 525 species of fish, invertebrates, birds, mammals and marine plans found in Monterey Bay are shown in 100 galleries and displays throughout the 2.2 acre (0.9 ha) site. Intake lines almost 1000 ft (300 m) long draw up to 2000 gallons (9000 litres) of sea water per minute from a depth of 55 ft (16.75 m) in the bay into the aquarium sea-water distribution system. The exhibits are carefully designed to represent the diverse ecological communities of Monterey Bay: deep and shallow waters, rocky and sandy intertidal zones, bays and estuaries, shale reefs, and the open ocean with its jellyfish, seals and whales. The point of this aquarium is to present a systems view of one place, unknown to most local residents and visitors except as a flat stretch of blue water.

Major exhibits

The three largest exhibits are the Kelp Forest (335 000 gallons/1.52 million litres), Monterey Bay habitats (326 000 gallons/1.48 million litres) and Sea Otters (55,000 gallons/250 000 litres). At 28 ft (8.5 m) high, the Kelp Forest is the tallest aquarium exhibit in the world, meeting the challenge of accurately recreating a living kelp forest community, complete with sardines, mackerel, a host of other fish, birds and many bottom-dwelling invertebrates. Four Bay Habitats have been reproduced in the next largest exhibit: the deep reefs, sandy sea floor, shale reefs and the wharf. This hourglass-shaped tank, 90 ft (27.4 m) long, allows large sharks, bat rays, ocean sunfish, salmon, striped bass and other open ocean fish to roam freely in full view of visitors. The Sea Otter tank is one of the most popular, since the small and endangered population of very endearing California sea otters has received much local and national attention.

Two other major exhibits are the Marine Mammals gallery, an overhead parade of full-sized replicas hung from the high ceiling, and the Great Tidepool, ringed on three sides by the aquarium and the fourth by artificial rocks that blend in with the surrounding granite boulders (Figure 9.5). Several displays provide opportunities for close encounters with marine animals – the tidepool touch tank, the bat ray petting pool, and the open air, walk-through aviary with its cross-section of the bay shoreline, from estuary to dunes to wave-swept beach. Here, snowy plovers, avocets, killdeer and other shorebirds scurry about at eye-level and almost within arm's reach.

The Monterey Bay Aquarium exhibits are an outstanding example of integrated design and consistency in text and message. The bioregional theme is stated throughout the aquarium, with example after example of marine plant and animal adaptations to specific habitats. Concepts are reinforced with video films, a hands-on microscope lab, introductory theatre presentation and a wall-size model of the Monterey Bay underwater region. Display copy is attention-catching, concise and accurate, reflecting writing standards that have been carefully thought out and followed. Perhaps the most outstanding interpretive feature of the exhibits are the many volunteer

Figure 9.5 Great Tide Pool at Monterey Bay Aquarium.

docents who are stationed throughout the aquarium, ready to answer questions and encourage interactive, personalized learning.

Education programmes

One of the most unusual aspects of the Monterey Bay Aquarium is its commitment to free education programmes and quorum admission for visiting school groups. Over 400 000 students have come from all over the state in the past 5 years to learn from more than 50 educational staff. To strengthen the effectiveness of school visits, the aquarium sponsors teacher institutes, workshops and lectures on marine science and environmental issues. Teachers are encouraged to plan ahead for their visit to the aquarium and are given free planning passes for admission.

The aquarium also hosts a creative outreach programme through the Aquaravan, bringing educational skills, classroom programmes and live marine life by travelling van to schools in the wider Monterey Bay region. The programme serves over 20 000 people each year, reaching out to special education classes, senior citizens in nursing homes and many thousands more at fairs, parks, libraries and community festivals. The Aquaravan has become famous among marine educators in central California for its ingenious costuming, interactive theatre and witty humour.

The education programme also trains and supervises more than 500 volunteers from the community who serve as aquarium guides. About 125 new guides are trained each year during a 48-hour, three-unit college course, in which they learn the exhibit floor and basic marine science concepts to share with the public. These guides are good-will ambassadors both to out-of-town visitors and back to the local community. They serve the important function of personal contact and presence in the midst of a large institution.

An additional arm of the education programme creates artistic and effective interpretive and classroom materials for teachers, aquarium members and the general public, some published in both English and Spanish to serve the local population. They also produce a series of natural history books about Monterey Bay and coastal California for school and public use. Education staff research, develop and write all the text for exhibit graphics as well as video scripts and training materials for aquarium guides. They have produced tour scripts in five languages and other brochures for visitors and are involved in planning and developing future exhibits, including the ambitious new wing devoted to the deep sea and outer waters of Monterey Bay.

Programme strengths

As a model for marine science education and interpretation, the Monterey Bay Aquarium has set a high standard for quality of exhibits and education programming. The regional emphasis establishes a clear message and conceptual framework for interpretation which is strategically reinforced through teacher training, curriculum and outreach programmes. The cumulative effect of repeated themes regarding local habitats is gradually building a community-wide understanding and appreciation of the local marine environment.

The original and consistent commitment to free education programmes and a full-scale education staff has made it possible for the aquarium truly to serve the local region and out-of-town visitors. Its materials have been distributed across the United States and around the world, serving as models for other aquarium-based marine education programmes. With a staff of over 50 educators, individuals are able to specialize in teacher training, guide management or outreach programmes. The education director is intimately involved in planning exhibit content and focus; thus, the exhibits have a consistent emphasis on education rather than entertainment.

The Monterey Bay Aquarium presents a model of coordinated education and interpretation, with generous programme follow-up and a diversity of possibilities for further learning. The education staff actively seek out schools and teachers to encourage science programmes based on field experience, and the administration wholeheartedly supports the extensive efforts of the education programme. The exhibits themselves stand as excellent examples of clear and evocative presentation of what are truly some of the most fascinating creatures of the world.

9.5 CONCLUSIONS

These three case studies present exciting possibilities for marine protected area interpretation in support of marine conservation and environmental understanding. They illustrate a diversity of effective techniques adaptable to the diversity of marine protected areas – each a unique combination of ecological habitats, management challenges and community resources calling for appropriate interpretation. Those beginning interpretive programmes can look to these examples and others for direction and models in developing public interest and awareness.

Education in marine protected areas is part of the challenge of expanding public awareness and appreciation of the ecological systems that support their lives. Marine conservation is a relatively new idea in many cultures, just as land conservation was a new idea for industrial countries more than 100 years ago. Marine parks and protected areas may some day be as common, as effective and as well-supported as terrestrial parks are today. Community involvement and public awareness can be the crucial links to building long-lasting and effective programmes. The beauty and diversity of marine habitats are drawing increasing public interest and concern. This attraction to the water is natural and powerful. Through effective education programmes, the marine environment may reveal insights and understanding not available from the land.

REFERENCES

Berkmuller, K. (1981) *Guidelines and Techniques for Environmental Interpretation*, University of Michigan, Ann Arbor, 98 pp.

Evans, J.G. (1988) *The Natural History of the Point Reyes Peninsula*, Point Reyes National Seashore Association, Point Reyes, 226 pp.

Gore, R. (1990) Between Monterey Tides. *National Geographic*, February, **177**, 2–43.

Kaza, S. (1985) *Education and Interpretation Plan for the Point Reyes–Farallon Island National Marine Sanctuary*, US National Marine Sanctuary Program, Washington DC, 172 pp.

Kaza, S. (1988) Community involvement in marine protected areas. *Oceanus*, **31**(1), 75–81.

Le Boeuf, B. and Kaza, S. (eds) (1982) *The Natural History of Año Nuevo*, Boxwood Press, Pacific Grove, California, 424 pp.

Sharpe, G.W. (1982) *Interpreting the Environment*, MacMillan, New York.

Strang, C. (1989) Project Ocean. *California Science Teachers Association Journal*, Summer, pp. 5–6.

Wood, D.S. and Wood, D.W. (1986) *How to Plan a Conservation Education Program*, US Fish and Wildife Service and International Institute for Environment and Development, Washington DC, 47 pp.

10

The voluntary approach to conservation of marine areas

Susan Gubbay and Sarah Welton

Most marine protected area (MPA) programmes are enshrined in legislation. Sites are designated under statute and management, or prohibition of activities within the protected area, is generally supported by additional regulations. A different approach, which has been practised in the Philippines and the UK for some years, is to rely on voluntary cooperation to protect sites. This chapter describes the contribution such initiatives can make to an MPA programme with particular reference to experience in the UK. Examples from the Philippines are described in more detail in Chapter 4, which looks at community involvement in both statutory and voluntary MPAs.

The value of encouraging people to support conservation measures of their own accord, rather than being required to do so, is widely recognized. It is a frequent theme of environmental education and interpretation (Chapter 9) and even enforcement programmes (Chapter 7). There is recognition that much can be achieved this way, but it is unusual to rely on voluntary cooperation to protect a site. The UK is one of the few countries to use this approach for much of its MPA programme. There are community-led voluntary MPAs, government-led voluntary MPAs, and situations where there is a statutory requirement to protect a site but where the government intends to rely on voluntary cooperation to achieve the protection.

10.1 THE VOLUNTARY APPROACH

Successful conservation management, whether for land or sea, depends on support and cooperation from users and from others who influence what happens in an area. In most cases cooperation is obtained through regulations and laws which deter inappropriate actions. However, no matter how many laws and wardens there are, those who do not agree with the regulations are likely to find ways of circumventing them. If, on the other hand, the majority of users favour a course of action, statutory powers may prove

Marine Protected Areas: principles and techniques for management. Edited by Susan Gubbay.
Published in 1995 by Chapman & Hall, London. ISBN 0 412 59450 1.

unnecessary. Voluntary MPAs aim to gain support for conservation without resorting to legislation. They rely on the cooperation of users and give the strong positive message that people can enjoy and use the area, providing that they do not threaten its well-being. This approach involves people who find legally based controls threatening or unacceptable, and helps to build support for marine conservation.

The aim of voluntary MPAs has been described as 'promoting harmonious use of the area, whilst preventing deterioration of the quality of the marine environment and the life it supports' (Gubbay, 1986). The emphasis is on using areas appropriately rather than on prohibition, even though there may be voluntary codes of practice asking people to leave certain areas undisturbed. In all cases a major concern is preventing deterioration of the marine environment, but the emphasis varies between sites. Some have an important educational role concentrating on interpreting the marine environment to the public. Others may be important for research, or for promoting actions which help to improve fisheries in the surrounding locality.

10.2 SUITABLE SITES FOR VOLUNTARY CONSERVATION

Experience with voluntary site protection in the UK and the Philippines has shown that this approach can be applied to many types of site. If an MPA programme is reliant on voluntary measures, the selected sites are likely to be areas of nature conservation importance and could include particularly diverse, fragile, vulnerable or productive areas. Voluntary codes of practice can be used to highlight the need for conservation of these areas and to give some protection in the absence of a statutory programme. They may act as a stop-gap in the hope that these key sites will have legal protection in the future. This view was taken in the UK in the early days of the MPA programme, and resulted in the establishment of **voluntary marine conservation areas** (VMCAs), two of which have since become statutory MPAs. Using VMCAs to protect key sites can, however, be a high risk strategy and most countries use a statutory MPA programme for these sorts of site.

Voluntary marine conservation areas can be set up to stimulate good conservation practice. The strong educational component and high level of community involvement in VMCAs can benefit any coastal locality and, because the idea is voluntary, it can and should be applied as widely as possible. A village beach or rocky reef which would not qualify for protection under a national programme could, for example, have an important role as a VMCA by focusing attention on the coast and encouraging people to get involved in marine conservation. Sites have become VMCAs because they are under threat, because the community has noticed a depletion of the marine flora and fauna, or because a local group is interested in taking the idea forward. They may be areas of local conservation interest, provide good opportunities for interpretation and education, or be popular areas for watersports. There is no need for detailed criteria to justify the choice of the site – if people are interested in setting up a VMCA they should be encouraged and supported to do so, regardless of the location.

10.3 ESTABLISHING A VOLUNTARY MARINE CONSERVATION AREA

Many aspects of the establishment and management of VMCAs are extremely flexible because they are not part of a formal programme. Action is led by local communities or specialist organizations. Different groups favour different approaches but there are a number of useful steps which should be considered. The most critical of these is to ensure that anyone who is interested can get involved, including those who do not support the idea. The participation of those who are uncertain about the value and role of a VMCA is very important, and their involvement can strengthen the group by ensuring that discussions and proposals are realistic and take account of all views. On many occasions those who are initially opposed to the idea become convinced of the value of a VMCA through their involvement and become its strongest advocates. Some opponents are easy to identify but there must also be a chance for those who are less vocal to participate, and for individuals as well as organizations to be represented.

A second important factor is the type of organization which leads the process. This is often a nature conservation organization but that need not be the case. Local community groups, schools, fishermen's cooperatives, research institutes and others have taken the lead in different places. Whatever the organization, it is important that it is considered to be an appropriate one, perhaps because of specialist expertise or because it is being impartial. There may be a change in the lead organization as the idea develops. For example, a conservation group could initiate the process and then a local authority might take the idea to the next stage.

A third critical success factor is patience. If all those interested in the area are to have a genuine opportunity to participate, it is inevitable that it will take time to set up a VMCA. The process may take months or years. Despite a natural desire to hurry it, those rushed into making decisions can become as negative about the proposals as they would have become if forced to take action by regulations.

The 10 years of experience in establishing VMCAs in the UK has shown a fairly consistent pattern (Figure 10.1). Generally a lead body or sponsoring organization, such as a local conservation group, university or local authority, first considers the possibility. Informal discussions lead to more formal meetings of interested groups and individuals to examine the feasibility of the concept, to consider the reasons for choosing the site and to discuss its

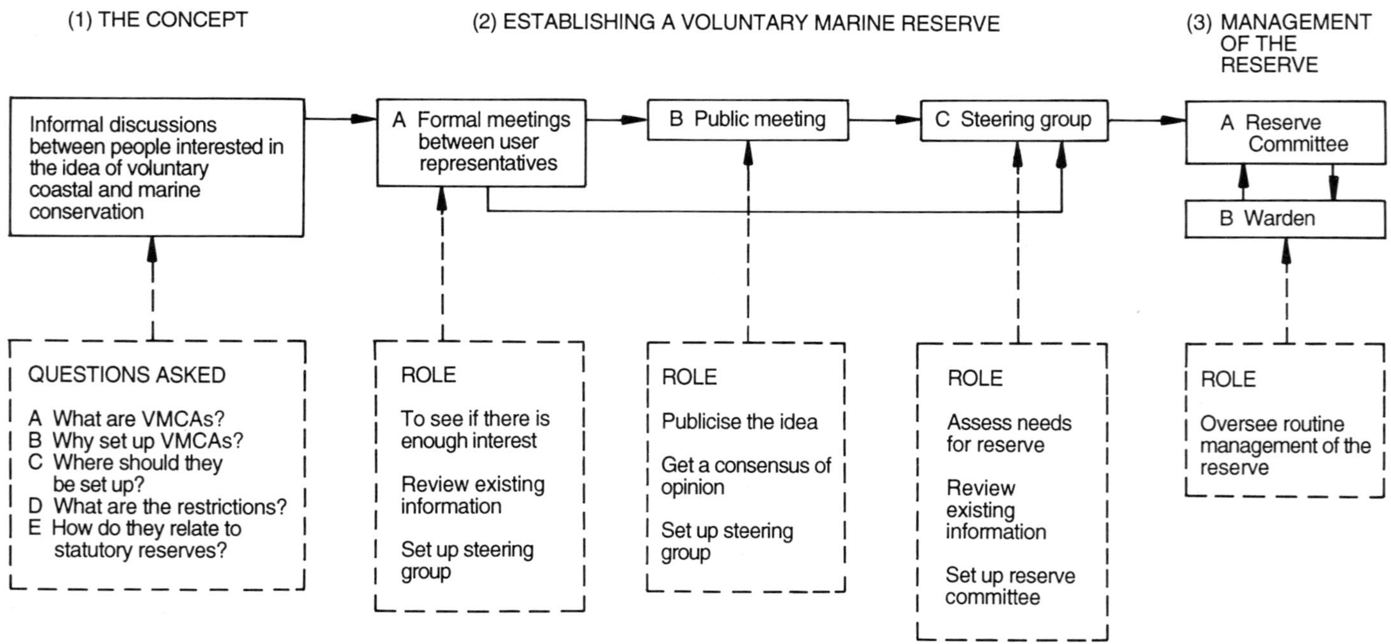

Figure 10.1 Procedures for establishing Voluntary Marine Conservation Areas (from Gubbay, 1986).

suitability as a VMCA. If there is enough interest and concern about the area, a steering group may be established to look formally into the possibilities of setting up such an area. Alternatively such a steering group may be set up after a public meeting discusses the possibility of a VMCA. Individuals on the steering group bring their specialist knowledge to the group, while sharing a common interest and concern for the area. They can pool their expertise in evaluating the area for its various interests and give initial thought to the aims of a VMCA and the activities affecting it.

Public meetings can also have a useful role in the process but can be a double-edged sword. They allow interested parties to voice their opinions, bring to light groups who were not represented at earlier meetings and present the idea of a VMCA to a wider audience. They do, however, need to be handled with care to ensure that the meetings are not dominated by vocal minorities. As VMCAs rely totally upon voluntary cooperation, it is of the utmost importance that everyone involved supports the concept. If there is strong opposition to the establishment of the voluntary conservation area, there is less chance of success.

Another way in which VMCAs have evolved is for a single group, such as a scuba-diving club using an area on a regular basis, to decide unilaterally on the introduction of a voluntary code of practice for its members' use of the area. This might include matters such as the taking of shellfish in the area. Other groups may follow suit and introduce their own codes to complement this action. As news of these codes spreads, other activities such as beach clean-ups or environmental days organized by local schools may be attracted to the area and a focus of interest builds up. This approach should be fostered, as it is the informality of VMCAs that encourages people to get involved and give support.

10.4 MANAGING VMCAS

Once agreement has been reached on setting up a VMCA, it is usual for a small group to take on the role of a reserve committee or management group. Such groups may be formalized through association with another body, such as a university or conservation organization, or may remain independent. The terms of reference and role of all management committees of this kind vary according to the aims and functions of the conservation area. They might simply have a watching brief or might be actively involved in producing management plans, promoting projects and providing interpretative material.

Management committees are likely to be involved in fund-raising. A voluntary organization, raising funds for a voluntary project, can often be more successful than a statutory body but such funds are generally raised on a year-to-year basis, resulting in an insecure future. It can be difficult to motivate a reserve warden or advisory group to produce expensive interpretative facilities at a site if activities might have to be curtailed the next year. Sponsoring organizations may also be reluctant to lay out funds on such an uncertain basis.

The benefits of a voluntary marine conservation area must be 'sold' to the user groups. Divers and snorkellers should come to regard a visit to the area as a privileged opportunity to explore the marine environment and should be encouraged to photograph and record marine life rather than collect it. Fishermen and anglers should see the site as a valuable breeding area for fish and shellfish, while shore-based visitors and educational groups should appreciate the value of the intertidal marine life for education and enjoyment (Figure 10.2). Either full- or part-time wardens are very important to this process.

At many sites in the UK, summer wardens have proved to be of great value in promoting an interest in the marine environment and in achieving conservation by voluntary cooperation. These seasonal wardens, generally marine biology students looking for experience in wardening, do an excellent job at these sites, especially in areas where there are large numbers of visitors. It is preferable, however, to have a full time warden or, if that is not possible, a long-standing volunteer with an interest in the day-to-day running of the marine conservation area. It takes time for a warden to

Figure 10.2 Group setting out for shore walk.

become familiar with the area and to form a trusting relationship with the local community and visitors. This relationship is vital if voluntary cooperation is to be achieved.

The responsibilities of the warden should include liaison with all those who use the area or have an interest in it, at all levels from the committee to the casual day visitor. The warden should attempt to become aware of the problems as (or, if possible, before) they arise. The warden can promote a code of conduct for the area, carry out 'on the ground' interpretation to the public and provide an advisory service for visitors, thus furthering the aims of the conservation area. Research and monitoring projects can be carried out in the area and the warden can encourage these to be undertaken by visiting groups. The impact of visitors can be reduced by guidance from wardens, who may also be able to persuade visiting groups to help with some of the site projects. A base, easily located by the public, from which the warden can operate can help fulfil these functions, and interpretative centres are often used for this purpose.

In achieving conservation by the voluntary approach, the way in which the warden presents the conservation area and its aims to visitors is of particular importance. With no legal backing, a warden cannot 'police' the area or enforce any of the aims. Success must be achieved by obtaining cooperation and by increasing public awareness of the marine environment and its needs. A long-standing warden who is trusted and respected by those who regularly use the area, and who has the support of the landowners and the local community, has a better chance of success. However, even full-time wardens cannot be at the site permanently. Local people can therefore be invaluable in supervising an interpretative centre in the warden's absence and in identifying and reporting any problems.

Formal and informal education play an important part in achieving voluntary conservation. Talks by wardens to school and college groups who visit the site not only help to get over a conservation message for the site but can also spark an interest in marine life and conservation which should last a lifetime. This can be extended into a visit to a seashore with guided walks, a service which is generally appreciated by teachers as well as being an enjoyable and memorable event for the children. Shore walks are an important part of the informal education of visitors of all ages. A well-presented walk, where there is a story to tell about the plants and animals found, is an excellent means of educating visitors to appreciate marine life.

Peer pressure is a particularly effective means of achieving conservation at a voluntary site. Once the site has become established and the aims well known, supportive members of user groups will influence others. Individuals who regard the conservation area as beneficial take on responsibility for its success. If they witness activities detrimental to the aims of the area, they will attempt to persuade the perpetrators that such actions are inappropriate. This means of voluntary control is often more effective than intervention by the warden but it usually only becomes apparent as a means of control once the area is well established.

10.5 EVALUATING THE SUCCESS OF VMCAS

The success of any conservation area in terms of the quality of the environment is always difficult to evaluate. In VMCAs, perhaps it should suffice that supporters believe there is an improvement. In areas where the emphasis is on education and interpretation, and the management of the area involves managing people rather than wildlife, the success can be evaluated in terms of the degree of cooperation and interest and the number of supporters. Voluntary marine conservation areas in the UK have brought publicity and recognition to sponsoring organizations and, in so doing, have increased awareness and concern for the marine environment. More tangible benefits have been the development of wardening programmes, establishment of interpretative centres and publication of booklets, leaflets and reserve guidebooks to provide information about the areas, as well as increasing the educational value and adding to the enjoyment of any visit to the coast. Voluntary codes of conduct have been drawn up to promote safety on the shore as well as to protect the marine life. Research programmes have been developed for a number of sites and in one case (the Helford River VMCA) a programme is under way to try to reintroduce seagrass to areas in which it was previously recorded (Thompsett and Turk, 1990).

Management plans have been produced for some VMCAs and the existence of these sites has played a part in establishing legislation for statutory marine conservation areas in the UK by providing working examples of how marine conservation areas can operate.

Another way in which VMCAs in the UK have been successful is by bringing people together to focus on the conservation of an area. The exchange of views and working together to resolve problems have been beneficial and have encouraged the formation of similar groups for larger areas of coast and even whole seas. A county-wide marine forum, organized on a similar basis to the VMCA management committees, was set up in Dorset following the success of the Purbeck Marine Wildlife Reserve (section 10.7). On a larger scale the North Sea Forum, later to become the Marine Forum, has extended the voluntary approach to cover a whole sea and represents international, commercial and recreational interests. Whether site-based or covering whole seas, these groups make a contribution to understanding and implementing wiser management of our activities in the marine environment.

10.6 THE UK EXPERIENCE WITH VOLUNTARY MARINE CONSERVATION

Early efforts to protect areas of marine nature conservation interest in UK waters grew from concerns expressed in the 1960s about threats to marine areas of biological importance. At the time there was no legislation which could be used to make these places marine protected areas, and no immediate prospect of an MPA programme. An alternative approach was needed and biologists working in the Bristol Channel, around Lundy Island, decided to promote a voluntary code of practice. With the support of the Lundy Field Society they established the Lundy Voluntary Marine Conservation Area in 1973. Other VMCAs followed and the momentum which they created for marine site protection, together with the efforts of those lobbying government, eventually led to a statutory MPA programme. The Wildlife and Countryside Act 1981 included provisions for the establishment of Marine Nature Reserves (Chapter 3) and a variety of options are now available for marine site protection (Table 10.1). Most of these options rely on voluntary measures and this makes the UK programme unusual in a global context and worth examining in some detail.

Three situations where a voluntary approach to site protection is being promoted are described below:

1. Voluntary MPAs which arc community led.
2. Voluntary MPAs which are government led.
3. Statutory MPAs which are largely reliant on voluntary measures.

There is most experience of community-led voluntary MPAs as they have been in existence since 1973. The other two ideas are relatively recent (1993 and 1994) but are already showing the benefits and problems which can arise when using voluntary measures under these circumstances.

10.6.1 Voluntary MPAs – community led

The British public have a great love for the coast and there are many examples where landowners and users have co-operated on a voluntary basis to maintain the quality of this environment. The extension of this idea to inshore waters through the establishment of VMCAs was perhaps a natural progression. There are currently 9 VMCAs (Figure 10.3) and voluntary site protection is now an accepted part of marine conservation programmes in UK waters. Acceptance has come for a variety of reasons. VMCAs have had an important role in educating the public to appreciate the sea and marine wildlife and to understand the need for marine conservation, and also in explaining how people can make a difference through their own actions on a visit to the coast. From the viewpoint of

Table 10.1 Management of sea areas in the UK for nature conservation (updated from Gubbay, 1993)

Title	*Year (of introduction or proposal)*	*Status*	*Responsible body*	*Number of sites*
Voluntary Marine Conservation Areas	1970	Voluntary	Locally based management groups	9
Marine Nature Reserves	1981	Statutory (Wildlife & Countryside Act, 1981)	Government conservation agencies[1]	2
Marine Consultation Areas (Scotland)	1986	Voluntary	Scottish Natural Heritage	29 (Scotland only)
Marine Protected Areas (Type 1 and Type 2)	1991	Proposal (statutory)	Marine Protected Areas Working Group[2]	–
Marine Consultation Areas (England and Wales)	1992	Proposal (voluntary) (withdrawn 1993)	Department of the Environment/Welsh Office	16 proposed (England and Wales)
Special Areas for Conservation (EC Habitats and Species Directive (92/43/EEC))	1992	Statutory. Regulation adopted 1994	Government agencies	–
Sensitive Marine Areas	1993	Proposal (voluntary)	English Nature	27 proposed

[1]Scottish Natural Heritage, English Nature, Countryside Council for Wales, Department of the Environment (N. Ireland), Joint Nature Conservation Committee.
[2]Group of conservation organization and individuals working on Marine Protected Areas.

Voluntary marine protected areas in the U.K.

1. St. Abbs and Eyemouth Voluntary Marine Conservation Area
2. Seven Sisters Voluntary Marine Conservation Area
3. Durlston Marine Research Area
4. Purbeck Marine Wildlife Reserve
5. Wembury Voluntary Marine Conservation Area
6. Helford Voluntary Marine Conservation Area
7. Isles of Scilly Marine Park
8. North Devon Voluntary Marine Conservation Area
9. Ceredigion Marine Heritage Coast

Figure 10.3 Voluntary Marine Conservation Areas around the UK.

national government the idea has particular appeal because there are likely to be fewer resource implications or commitments in a busy legislative timetable.

The detailed arrangements of VMCAs vary but they tend to follow the pattern described in section 10.3. Although none of the sites have any legal status there may be some statutory controls, such as local authority or sea fishery committee by-laws, which are not linked to the establishment of the VMCA but can help with the management of the site. Codes of practice and wardening are important ways of promoting sensitive use of the site, but if people choose to abuse the aims of the conservation area, those promoting the voluntary codes are powerless to intervene. Total cooperation is therefore the key to success and although this may seem unrealistic, a great deal has been achieved at many of the VMCAs. This is illustrated by the case of the Purbeck Marine Wildlife Reserve (section 10.7). The continuing momentum for the establishment of VMCAs, with the most recent site being agreed in 1994 (Table 10.2), shows that interest in this approach remains strong more than 20 years after the establishment of the first VMCA.

10.6.2 Voluntary site protection supported by government

In the UK, the statutory programme of site protection has never caught up with voluntary initiatives. Two **marine nature reserves** (MNRs) have been

Table 10.2 Voluntary Marine Conservation Areas in the UK

Year	*Name*	*Notes*
1973	Lundy Island	Became a statutory reserve in 1986
1974	Skomer Island Marine Reserve	Became a statutory reserve in 1990
1978	Purbeck Marine Wildlife Reserve	
1981	Wembury Marine Conservation Area	
1982	Roseland VMCA	Limited activity in recent years
1984	St Abbs and Eyemouth Voluntary Marine Reserve	
1987	Helford River VMCA	
1987	Seven Sisters VMCA	
1989	Isle of Scilly Marine Park	Set up after proposals for a statutory reserve were rejected
1992	Ceredigan Marine Heritage Coast	
1994	North Devon Marine Conservation Area	
1994	Durlston Marine Research Area	

declared since 1981 and discussions over the designation of MNRs on the Menai Straits in North Wales and Strangford Lough in Northern Ireland have been dragging on for more than 10 years with no end in sight. Plans to designate the remaining three sites from an initial list of seven have been put 'on ice' and there are no signs that new sites will be added to the list for consideration. There has been a distinct shift away from the statutory programme and the UK government is now supporting and developing voluntary initiatives (Gubbay, 1993b).

One area where this shift can be seen is in the recognition and support that government agencies are giving to VMCAs, which are being highlighted in local plans and coastal conservation strategies. In 1993 English Nature, the statutory advisers to the government on conservation matters in England, set up a grant fund to support VMCAs. The fund has been particularly helpful because financial insecurity is common to most VMCAs. In its first year money from the fund has been used to support surveys and monitoring, the preparation of educational and interpretation materials and an updated guide to setting up VMCAs (Welton, in press) and to assist with the costs of liaison work and running meetings. The range of projects, and the fact that the fund is already oversubscribed, shows the value of this sort of support.

As well as helping the voluntary initiatives of others, the government agencies are developing their own voluntary programmes. The 'Firths Initiative' of Scottish Natural Heritage, which promotes coastal zone management, and English Nature's Sensitive Marine Areas programme are two examples. In 1993 English Nature published its strategy for marine conservation, in which one of the stated aims was to establish a framework for cooperative management of key marine wildlife areas. It hopes to achieve this by working with others to develop marine management techniques and involving the community (English Nature, 1993a). Twenty-seven sites have been identified as 'sensitive marine areas' (English Nature, 1993b) and English Nature sees its role as supporting those areas which have existing initiatives by providing specialist information to help decision-making and the development of management plans. In the remainder it plans to work with those who are responsible for management, or who use the area on a regular basis, to develop voluntary management options which can supplement existing regulatory controls.

The promotion of a voluntary initiative by an agency set up to advise government and run the statutory MPA programme led to some confusion. Although the documents stated very clearly that English Nature would seek voluntary agreements for the management of these areas, there was suspicion that management agreements relating to these sites were not really going to be voluntary because the idea was being promoted by one of the bodies responsible for the statutory programme. Additional confusion has been created by the introduction of a European Union Directive because many of the sites which will qualify for protection under the terms of the Directive are 'sensitive marine areas' identified by English Nature's voluntary programme. The sensitive marine areas programme is in its early days and, as yet, there are no management agreements. Whatever happens, if this approach is taken by government agencies or statutory advisors it is essential to make a clear distinction between the different approaches being taken by the same agency.

The combined approach of voluntary initiatives with government backing has a great deal of potential and is discussed further in Chapter 4 using the example of community-run voluntary reserves in the Philippines. Caution is required, however, in relation to the amount of government involvement. If there is too much it will be perceived as being government led and therefore not truly voluntary, whereas too little could be seen as government abandoning community initiatives.

10.6.3 Statutory MPAs which rely on voluntary protection

The most unusual use of the voluntary approach in the UK is to help to fulfil a legal requirement, using statutory powers as a last resort. This situation has arisen with the UK's plans for the implementation of the EC Directive on the Conservation of Natural Habitats and of Wild Fauna and Flora (92/43/EEC, the 'Habitats Directive').

The aim of the Directive is 'to contribute towards ensuring bio-diversity through the conservation of natural habitats of wild fauna and flora in the European territory of the Member States to which the Treaty applies' (Article 2.1). It calls for the setting up of a coherent European ecological network of **special areas of conservation** (SACs) for habitats and species listed in annexes to the Directive. Marine habitats and species are included on these lists, hence the need for MPAs to fulfil the requirements of the Directive. The UK government proposes that, in the case of marine SACs, organizations should work together and draw up management plans through voluntary agreement but with no single body having overall authority. If this approach fails, government ministers will have statutory powers to intervene. The emphasis is therefore on voluntary cooperation, but with a back-up system if this fails (DoE/Welsh Office/Scottish Office, 1993).

These proposals have met with strong criticism from non-governmental organizations, who consider that benefits which come from taking a voluntary approach, such as good will, and the absence of any requirement to participate will not apply. On the other hand, there is to be no clear direction, through statutory measures, unless the voluntary approach fails. The result could be described as the worst of both worlds.

10.7 THE PURBECK MARINE WILDLIFE RESERVE: A CASE STUDY

The Purbeck Marine Wildlife Reserve became a VMCA in 1978 under the guidance of the Dorset Wildlife Trust. The reserve is on the south coast of the Isle of Purbeck, between the popular south coast holiday resorts of Swanage and Lulworth Cove, and includes 8 km of the south Purbeck coast cliffs and beaches. The seaward boundary extends approximately 1 km offshore to a maximum depth of 20 m (Figure 10.4). The shores show a wide range of exposures to wave action with hard rock ledges acting as natural breakwaters. These ledges run a considerable distance out to sea as a sublittoral wave-cut platform, and provide horizontal and vertical surfaces for colonization by marine life.

The VMCA was set up because of visitor pressure on the area, the nature conservation and geological interest of the site and the potential for educational use. The Smedmore Estate owns Kimmeridge Bay – the main access point to the reserve – and the beach down to low-water mark known as 'Lord of the Foreshore'. This control over access through ownership and the interest of the landowner in conserving the area have had a great influence on the success of the VMCA.

The western part of the reserve is controlled by the Ministry of Defence and is inaccessible to the public at most times. This area is in frequent use as a tank firing range but the large outer danger area, established for safety reasons, means that some 8 km (5 miles) of coast are virtually undisturbed. In contrast, the shores in the eastern section of the reserve are under increasing pressure from holiday-makers, and from recreational and educational groups, as more than 100 000 people visit Kimmeridge Bay each year. Too great an intensity of use may destroy the very features of the area that are in demand, and the need for quiet recreation cannot be reconciled with noisy activities. Consequently there is a policy of encouraging selected activities and discouraging others so as to maintain and, where possible, enhance the value of the conservation area. Advertising the existence of the VMCA has required careful handling since declaration of a conservation area will inevitably lead to some short-term increase in interest. Long-term policy has been to acquaint visitors with the existence of the VMCA once they have arrived via information panels, leaflets, visitor centres and wardens.

10.7.1 Setting up the VMCA

At the initial public meeting when the idea of a voluntary conservation area was presented by a member of the Dorset Trust for Nature Conservation (now known as the Dorset Wildlife Trust), the voluntary status of the proposed conservation area was stressed. The meeting was attended by representatives of user groups and, following discussions on the aims and

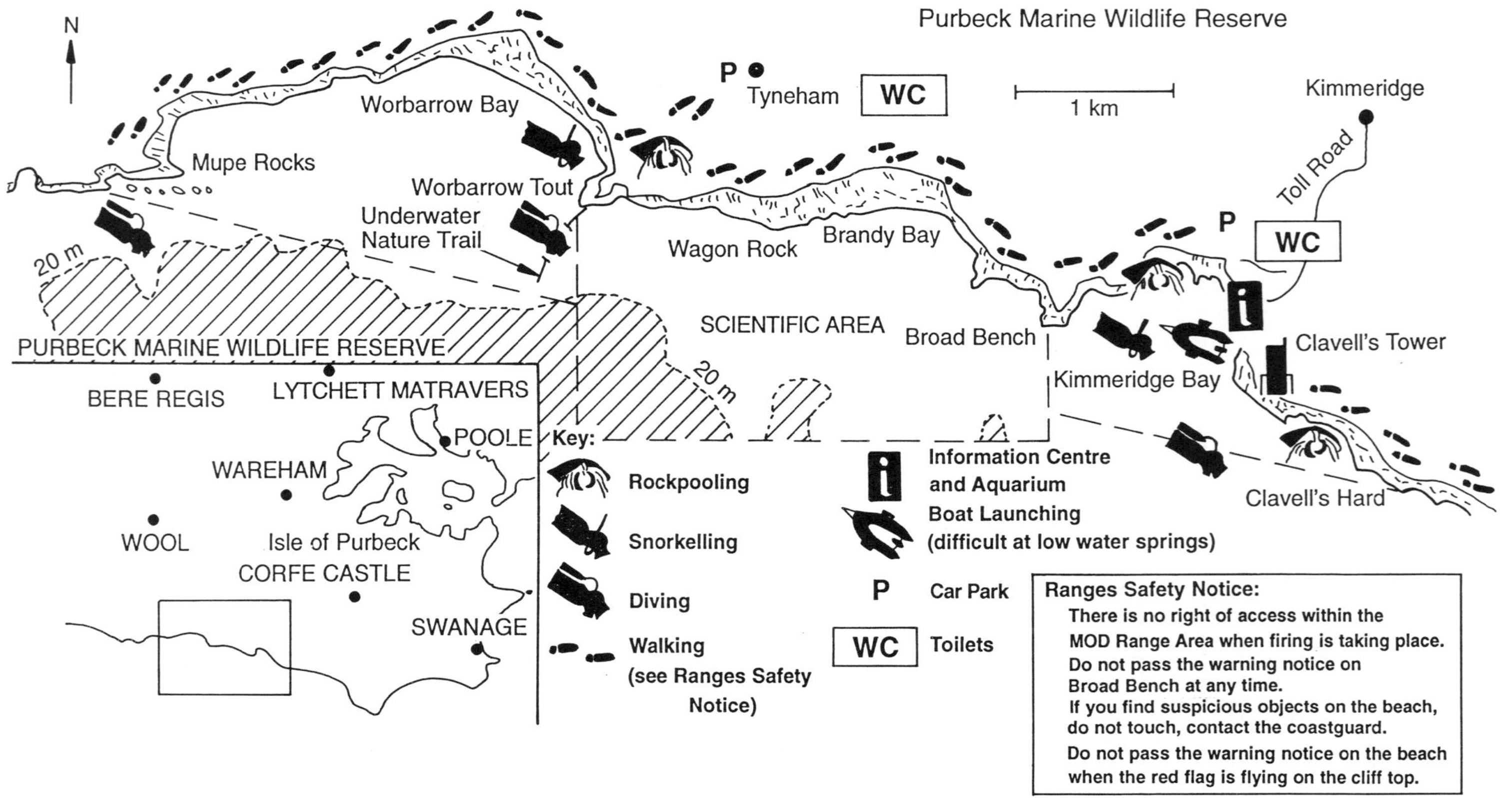

Figure 10.4 Purbeck Marine Wildlife Reserve.

objectives of the proposed conservation area, everyone supported the idea. Without such total support there would have been no Purbeck Marine Wildlife Reserve.

The next step was to form an advisory group and all organizations with an interest in the area were invited to send representatives. Its functions are:

- to encourage achievement of the reserve's objectives;
- to keep the reserve guide up-to-date;
- to stimulate production of educational materials in support of the reserve for schools, college parties and members of the public; to coordinate research and surveillance;
- to inform represented organizations of the progress of the VMCA and of action required from them to support the conservation of the area.

The advisory group prepared a reserve guide (Dorset Naturalists Trust, 1978) which described the value of the area in nature conservation terms and listed the activities which took place as well as the potential threats. It also outlined the aims and objectives of the VMCA, a general code of practice for the area and specific codes of conduct for divers and seafishermen (Appendix 10.A).

The advisory group has met twice a year since 1978 and has supervised the management of the reserve. More than 20 organizations are represented and members have been enthusiastic and supportive. They have found meetings an excellent opportunity to discuss all aspects of coastal conservation in an informal, voluntary atmosphere. An annual report of the advisory group, funded by British Petroleum, has proved a useful means of explaining the management of the reserve to individuals, organizations and potential sponsors.

10.7.2 Wardening and interpretation

There was some doubt about whether the voluntary approach would be effective and the next step was to be the real test. The idea had certainly proved successful in the case of the advisory group, but there was a need to work with the public. Funding was obtained to employ a full-time reserve warden in 1979 with the initial task of identifying users of the area and informing them of the existence of the reserve and its aims. The decision was made to 'tread carefully' as few visitors were likely to understand what a marine reserve was and there was little back-up information on the site to enable them to learn more. This changed after the Dorset Wildlife Trust acquired a fisherman's hut next to the boat slipway at Kimmeridge Bay. This became the Kimmeridge Information Centre, where a simple display was installed illustrating the geology, history and wildlife of the area. As funds became available the centre was made more appealing with the introduction of lighting, an aquarium, a touch shelf and an area for children's drawings and work. Interpretative material has been produced to enhance the enjoyment of visitors and to promote the aims of the reserve. Leaflets and nature trail guides are sold from the centre and marine life colouring pictures are very popular with the younger visitors. The centre is also a base for the warden and has become the focal point of the reserve. The warden can meet organized groups at the site and lead guided walks from the centre. People visit it because they wish to understand more about the area and this also makes it an ideal place to meet people and tell them about the reserve, its aims and how they can help (Figure 10.5).

Following the success of the Kimmeridge Information Centre, similar centres have been established at Tyneham Church in the Ranges and in nearby Lulworth Cove. Although Lulworth is not in the reserve, it is the next accessible rocky shore and is a popular boat-launching site for divers visiting the reserve. The centre at Lulworth is a small building adjacent to the pedestrian road to the beach and has provided an excellent base for summer wardens. The landowner, sympathetic to the voluntary cause, provided the Trust with the building and materials for its upkeep. In the absence of seasonal wardens, the running of this centre is supervised by the local community. It provides an excellent opportunity to get the message

Figure 10.5 Visitors to the Kimmeridge Information Centre.

across to many more people as Lulworth has around half a million visitors each year.

The voluntary status of the Dorset Wildlife Trust and the conservation area has helped to attract financial and material support for wardening. In addition to financial sponsorship of the warden's post, equipment and administrative assistance have been forthcoming. A caravan was donated to accommodate assistant wardens and planning permission was successfully sought to place the caravan in a local site. This facility has made it possible to use volunteer student help in an expensive holiday area.

10.7.3 School groups

In the Purbeck Marine Wildlife Reserve, management centres around education and encouraging quiet enjoyment of the area and its wildlife. The warden and seasonal assistant wardens meet around 200 school parties each year. Older groups of geology, geography and biology students undertake field work required by school syllabuses. Although study is generally planned by the teacher in question, the visits provide an opportunity to explain the voluntary nature of the reserve, how the area is managed and how conservation is achieved.

The majority of the educational groups who visit the reserve are between 9 and 12 years old. The wardens encourage these groups (many of whom have little opportunity to visit seashores) to explore, discover and enjoy the wealth of wildlife in the reserve. Every child enjoys exploring the seashore and with guidance can learn how to discover and enjoy marine life without harming it. An introductory talk is given by the wardens. The geology and history of the area are explained – smuggling stories and the Kimmeridge ghost always go down well! Simple principles of conservation, such as replacing stones and collecting only empty shells, are incorporated into stories about the plants and animals that they will find on the beach. Once on the beach, children are encouraged to imagine what it is like to live there and how they would survive storms and the effects of the sun and the wind if they were a barnacle or a sea anemone. Children will be considerably less likely to kick limpets off rocks if they are asked to consider what it is like being a limpet.

The success of this informal approach with visiting school parties can be evaluated from the excellent response from children on the beach and the work that is later sent to the wardens. It has also shown how children can educate their parents. Children from as far afield as London and the Midlands, who have visited with a school party, frequently then bring their parents to the reserve in the holidays. It is gratifying to discover how much they have remembered and understood and how interested they are.

10.7.4 Promoting the voluntary conservation message

In addition to educational parties, large numbers of holiday-makers visit throughout the year. The wardens concentrate on trying to meet as many visitors as possible, in the Information Centre and on the beach, since explaining the importance of the reserve, its voluntary status and the need for conservation has proved most effective on a one-to-one basis. Visitors feel that they are being consulted about the voluntary reserve and have some input into it and therefore have a vested interest in the success of the project. When the wardens are not available the Information Centre, information panels and publications promote the reserve and its aims (Figure 10.6).

Guided walks have proved to be an excellent means of getting the message across. On 'Rock Pool Rambles' or 'Seashore Saunters', adult visitors are welcomed to the reserve at a low tide and are shown the wealth of wildlife on the shore. 'Rocks and Fossils' walks explain the geology of the area and show visitors how they can find fossils safely. 'Rock Pool Adventures' and games for children are extremely popular, and thousands of children and their parents take part each year. Owing to the popularity of these events (more than 120 participants on one occasion), knowledgeable volunteers and local wardens are brought in to help. The walks are fun and the

Figure 10.6 Voluntary code of practice.

children are being educated without realizing it. Following the success of these children's events, similar walks are now organized at other sites on the Dorset coast. As well as the walks, guided boat trips are run every Saturday morning in the summer. The wardens explain the geology of the cliffs and the historical landscape to the passengers. These trips are also an opportunity to explain to non-divers what lies beneath the waves. In calm weather, seaweeds and animals can be seen on the sea-bed, and passing jellyfish and shoals of mackerel are a bonus. A plankton trawl always amazes people when they examine the 'clear water' and see that it is alive with microscopic creatures. Boat trips are one of the best ways to explain why the reserve exists and how it works. Visitors can feel that they are part of the marine environment and the extent and boundaries of the reserve can be shown.

Wardening at the Purbeck Marine Wildlife Reserve encourages sensible, non-destructive use of the shore for both educational and recreational purposes. Collection of live specimens and intentional damage are actively discouraged but, inevitably, in a voluntary conservation area with limited wardening resources, these do occur. As the voluntary reserve status has become more widely known, these events have become less frequent. At Kimmeridge Bay, in the early days of the reserve, local reductions in abundance of limpets and winkles were noted, together with noticeable changes in population size and structure due to human predation upon the largest individuals. Some attractive species of algae also proved vulnerable to collection, particularly the peacock's tail weed, *Padina pavonia*, and the colourful weed *Cystoseira tamariscifolia*. As visitors have become aware of the concept of the voluntary conservation area, peer pressure has helped to reduce collection, intentional damage and damage through ignorance. Children on a guided walk or on an educational visit, who have just had the principles of conservation explained to them, will point out to other children on the beach the advisability of returning crabs to the sea and of checking whether shells are empty. A 7-year-old on a school visit was overheard explaining to a toddler that anemones did not like having stones piled on top of them: he told the child that anemones were living creatures and the stones might hurt. This type of general public education is one of the positive ways of minimizing damage within such an area.

The effects of incidental trampling are obviously considerable and changes in species abundance of intertidal communities have been noted. A study of the effect of trampling upon marine communities in Kimmeridge Bay was carried out (Fowler, 1981) and basic monitoring in subsequent years demonstrated the effect of trampling on the foreshore. The only way to control this damage is by managing the intensity of use of the shore, and the most suitable way of doing so is through access controls. In the Purbeck Marine Wildlife Reserve access is limited to Kimmeridge Bay and is thus managed automatically. Levels of use within the area are so high that this situation represents an ideal management regime, the Bay being 'sacrificed' as the centre of interpretation and maximum use, whilst the more remote parts of the reserve are maintained in an undisturbed state. Access and navigation within much of the reserve are restricted by the Ministry of Defence since it falls within part of the Lulworth Ranges Sea Danger Area. This provides a relatively undisturbed section of the reserve which has been defined as a scientific area.

On shore, access to the ranges is confined to weekends and holiday periods and walkers must not stray from the footpaths. This section of the reserve provides an ideal control, when compared with Kimmeridge Bay, to assess the effect of visitor pressure. Scientific research is encouraged within this area. Boat users are asked to avoid the scientific area as far as possible, and to take particular care below Gad Cliff during the bird-nesting season.

10.7.5 Watersports

The reserve area is extremely popular for all forms of watersports. Surfers, canoeists and windsurfers take advantage of the surf waves which roll in over the wave-cut platforms. Very little management of these groups has been required because it is felt that their activities do not pose a threat to the reserve and they generally visit in weather conditions unsuitable for other beach activities. Jet-skis, on the other hand, which recently appeared in Kimmeridge Bay, were felt to be inappropriate in the reserve. These machines were ridden in fine, calm weather in the summer, stirring up the sediment in shallow water, posing a danger to bathers and snorkellers and creating a great deal of noise. As one of the main aims of the reserve is to encourage quiet enjoyment of the marine environment, the landowner banned jet-skis from the area. Although the reserve is voluntary, the access road is privately owned and with the cooperation of the landowner such bans can be enforced.

The reserve is one of the most popular scuba-diving sites in the country and is visited by diving clubs from all over the UK. Boat launching facilities are good and the varied substrata and wrecks provide excellent dive sites. Although diving clubs can launch from Lulworth Cove to dive in the reserve, most launch from the slipway at Kimmeridge Bay. This helps wardening, as divers can be met both as they prepare to launch and after their dive.

The voluntary approach can be particularly effective with divers as they can be convinced that the existence of the reserve is in their interest. They all want to see colourful marine life on their dives, and indeed many diving clubs visit the area because it is a marine nature reserve. Peer pressure plays an important part when promoting the aims of the reserve to divers. If a diving club has one member who supports the reserve, that person will influence other members. Continuity of wardening has also proved to be important when dealing with divers. It takes several years for a warden to get to know the divers who visit regularly and to gain their trust. The Spear-fishing Association has agreed that its members will not fish in the reserve.

The provision of diving information provides an opportunity for the warden to get to know visiting divers. Charts and tidal information are available, and marine life photographs aid identification. A permanent underwater nature trail extends from the shore to a depth of 15 m, crossing several different habitats and guaranteeing a varied dive. A rope bolted to the rocks guides visiting divers, and numbered stations and labels explain the wildlife along and near the trail. A visitors' book (a slate and pencil) provides an amusing touch. This facility aims to encourage divers to appreciate the marine life they see and is very popular with diving clubs, especially those training novice divers. The effect of diver pressure on the trail has been studied and compared with a similar trail, known only to the survey team (Collins and Mallinson, 1989).

Angling is welcomed in most of the reserve. Anglers are asked to respect the aims of the scientific area and agreement has been reached that the area will not be used in competitions. Anglers are also asked to help research by keeping records of their catches in the reserve and to follow the practices recommended by the National Federation of Sea Anglers.

10.7.6 Commercial activity

The current level of commercial fishery in the VMCA, mainly for crab and lobster, is not regarded as a threat. This is a long-standing activity and can provide valuable information on natural processes and fish population. The sea-bed in the reserve is not suitable for trawling or dredging but any development of new intensive methods may pose a threat to the reserve policy. A reserve code of practice states that in order to help to identify long-term biological changes, the reserve advisory group would welcome records of catches within the reserve. Such records may be placed with the Dorset Environmental Records Centre, with the usual arrangements for confidentiality. Should there be evidence that an activity is causing or is likely to cause harm to fish stocks or their environment, fishermen would be informed. The voluntary nature of the reserve has resulted in an excellent working relationship between the warden and local fishermen and their cooperation is sought if problems arise.

The situation with the oil and gas industry, on the other hand, has not been favourable to the VMCA. A block off the south Purbeck coast, encompassing part of the reserve, was allocated in a recent oil and gas licensing round. The oil company concerned has been carrying out environmental studies and is planning seismic activity in consultation with local authorities and the statutory agencies. The reserve's advisory group has not, to date, been included in this consultation. Activities such as oil exploration, which could have a profound effect on the reserve, can take place in or adjacent to the area without requiring consultation with the reserve management, owing to its voluntary status. The amount of consultation will, instead, depend on the oil company concerned.

10.7.7 Implications beyond the reserve boundary

The voluntary approach to managing the reserve extends far beyond its boundaries. Illustrated talks about the reserve and its wildlife are given by the warden to local and national organizations, schools and diving clubs. Many local organizations follow up a talk with a visit to the reserve. As this is one of the first VMCAs in the UK and the first county conservation trust to employ a full-time marine member of staff, the warden is frequently requested to speak at meetings and conferences and to present the voluntary approach to other coastal trusts and organizations. The Dorset Wildlife Trust finds that the Purbeck Marine Wildlife Reserve has brought the Trust considerable recognition and publicity. It has also attracted sponsorship and has been virtually self-financing.

In 1987 members of the reserve's advisory group agreed that meetings were an excellent voluntary forum at which to discuss issues about the VMCA and other parts of the Dorset coast. It was felt that the group should therefore extend its remit to the whole Dorset coast. To attract increased sponsorship to secure the future of the VMCA, it was also decided that a marine committee of the Dorset Wildlife Trust be formed to give the advisory group some official status. The committee has a large membership of organizations with an interest in the Dorset coast which makes it difficult to have regular meetings but an annual seminar is held for representatives of all member organizations, on a subject chosen by the executive sub-committee. Reports of the seminar proceedings are published by the Dorset Wildlife Trust and sent to members, libraries, schools and colleges. The aims of these annual seminars are to stimulate thought and discussion and to encourage continuing dialogue among all the many interests represented. The smaller group meets regularly to deal with any issues concerning the conservation of the Dorset coast.

Some of the most enthusiastic representatives on the marine committee wanted nothing to do with conservation when they were first approached. The voluntary approach, where a Harbour Master is asked for his advice over conservation of the harbour, was soon found to be a different matter from a conservationist telling him what he should do with his harbour!

Although the success of the marine committee and the Purbeck Marine

Wildlife Reserve is difficult to evaluate, the increasing support and cooperation from committee members and the public suggests that the voluntary approach is working. A high level of cooperation has been achieved by the work of the wardens over many years. The objectives of the reserve, outlined in the reserve guide, are being achieved while retaining the voluntary nature of the project.

10.8 CONCLUSIONS

Traditional rights, freedoms and open access to much of the marine environment makes attempts to control or prohibit activities in MPAs a complex issue. Because of this, the idea of protecting sites through voluntary means can be appealing.

Experience in the UK with the establishment of VMCAs has shown that it can be a very positive process. People feel they can make a real contribution to marine conservation in their locality and much can be achieved in promoting the message of marine conservation, more generally, through education and interpretation. Progress can also be made with site protection if there is no possibility of statutory control. At all the sites which have been established in the UK it is the recreational and educational activities that are covered by voluntary codes of practice. Commercial activities are unaffected, except through the controls imposed by their own industry which may not be specific to the VMCA. The sites are getting more status, however, by being recognized in local development plans and guidance notes and it may be that, in the long term, industries will make some distinction between their operations inside and outside VMCAs.

The most significant drawback of voluntary protection for MPAs is that there is no safeguard if the site is under threat. Wilson (1991), in his review of voluntary sites, noted that these areas are generally effective against minor threats but are unlikely to compete with major problems and this is amply illustrated by the threat of oil and gas exploration adjacent to the Purbeck Marine Wildlife Reserve. Statutory protection does not always guarantee protection but a stronger case usually has to be made to overrule a statutory MPA. In the case of VMCAs it is most often peer pressure and the success of education and interpretation programmes that determine how well a site can be protected and this can fail when weighed against economic realities. Control of commercial fisheries, or oil and gas extraction, in a VMCA has been hard to achieve and there is a danger that this can undermine the credibility of the whole approach. Problems with continuity of support, either financial or in personnel, are also difficulties for VMCAs.

The success of all MPAs is linked closely to community and user support but in the case of VMCAs, where there is no legislation to fall back on, there is generally a heightened awareness of the importance of getting people involved. This can be seen from the outset with local users suggesting that an area should become a VMCA, whereas experience of most statutory programmes is that government agencies make the initial proposals and then seek discussion with interested parties. Another very important effect of the absence of legislation is that people are under less pressure to cooperate or even to be involved. The choice is theirs. One outcome is that people feel more motivated to participate and this creates a positive feel in the process of site identification and management rather than the MPA being seen as a response to a legal requirement. There is a danger, however, that this support can be undermined if voluntary protection is used as a stop-gap on the way to statutory designation. This could cause problems when other potential VMCAs are being considered, particularly if people perceive them as a step on the way to statutory controls.

The emphasis on education and interpretation is another strength of VMCAs. This can be done regardless of the legal status of a site and has proved to be a key objective of VMCAs. It is a strong feature of all the sites and although its value is difficult to quantify, the promotion of good practice on the seashore and at sea is always worthwhile.

Critical factors which determine the success of sites are, first and foremost, to involve as many interested individuals and organizations as possible. It is especially important to draw in those who are likely to be objectors, from the start. This will help to ensure that the discussions and proposals are realistic and robust. The need for continuity is also critical. VMCAs are often driven by key individuals who are especially enthusiastic and give their time to the project. A great deal can be achieved by these individuals but it is also essential to have some form of management committee as personnel and circumstances change. A group such as this should be set up as early as possible. A related factor which influences the continuity of a VMCA is financial security. As these areas are outside a statutory programme there is generally no automatic means of funding. This is one of the major difficulties affecting sites and ways around it have been sought through sponsorship, help in kind, etc. Early consideration of the financial situation is therefore important but this should not preclude taking action as experience has also shown that a great deal can be achieved with little money through the enthusiasm of those involved.

Finally, an important message is to be patient. It can take many years for a VMCA to become established or it may be that proposals for some sites are rejected by the community. The most important factor for those involved is not to force the situation. If it is forced, it is not voluntary and a lot can be lost.

The voluntary approach to marine site protection is not suitable for all localities. Key sites should be given statutory protection but there are many lessons from the voluntary approach which are useful for statutory sites.

The most significant of these is the importance of encouraging the involvement of people who live, work or play in the area. The voluntary approach has an extra dimension as well and that is to promote marine conservation at any locality, regardless of its marine biological importance. Much has been achieved, with limited resources, as the voluntary approach has often harnessed people's enthusiasm and commitment because they feel that they can make a personal contribution to marine conservation in their locality.

APPENDIX 10.A SEASHORE AND UNDERWATER CODES

Seashore Code

Enjoy these beaches and please leave them so that others may enjoy them.

The following suggestions are intended to help preserve the beauty and interest of the shore.

PLEASE . . .

- Remember that wild plants on cliffs are easily dislodged.
- Avoid uprooting or trampling clifftop plants.
- The animals that live on the underside of the rocks soon die if left exposed.
- When searching among rocks, replace stones you move as you found them.
- Most sea animals are delicate and easily damaged.
- Only handle sea animals if there is a good reasons for so doing; take notes, make drawings, take photographs.
- Do not collect without good reason. If you do collect:
  - There is rarely a need to collect live animals. Empty shells don't smell. Collect dead shells.
  - Never collect anything that appears to be scarce.
  - Collect sparingly. Leave plenty for others to enjoy – including fossils, remember they have been in the rocks for many millions of years.
- Do not leave litter, fires burning or spill liquids on to the beach or into rock pools.
- Educational Groups: Regular monitoring is essential to coastal conservation. Please send a summary of your records to: The Dorset Environmental Records Centre.

Underwater Code

The following suggestions are intended to help conserve underwater life so that it remains unspoilt for everyone to enjoy.

- Fish can be made shy by spear-fishing and other people's enjoyment is thus spoilt. Please do not spear-fish.
- Change of position disturbs sedentary organisms which then may die.
- Please leave apparently stable rocks as they are.
- Many sea organisms take years to grow and populations may be slow to recover after severe conditions have damaged them. Some growths may take twenty years to reach attractive size.
- Please do not collect Sea Urchins, Sea Fans or other attached animal growths.
- If you collect, please do so sparingly and avoid taking live organisms unless there is a special reason.
- Please do not leave litter or spill liquids.

Accurate notes, drawings and photographs of underwater life, supplied with date, detailed location and local conditions, will be gratefully added to County Records if sent to: The Dorset Environmental Records Centre.

REFERENCES

Collins, K.J. and Mallinson, J. (1989) *The Kimmeridge Underwater Nature Trail Project*, Dorset Trust for Nature Conservation.

Department of the Environment/Welsh Office/Scottish Office (1993) *Implementation in Great Britain of the Council Directive on the Conservation of Natural Habitats and of Wild Fauna and Flora (92/43/EEC), 'The Habitats Directive'*, Consultation Paper.

Dorset Naturalists Trust (1978) *Purbeck Marine Wildlife Reserve*, Reserve Guide.

English Nature (1993a) *Conserving England's Marine Heritage. A strategy*, English Nature, Peterborough, 20 pp.

English Nature (1993b) *Important areas for marine wildlife around England*, English Nature, Peterborough, 156 pp.

Fowler, S.L. (1981) *The Ecological Effects of Visitor Pressure on the Rocky Inter-Tidal Zone of the Purbeck Marine Wildlife Reserve, Dorset*. MSc dissertation, University College, London.

Gubbay, S. (1986) *Conservation of marine sites. A voluntary approach*, Marine Conservation Society, Ross-on-Wye.

Gubbay, S. (1993a) Management of Marine Protected Areas in the UK: lessons from statutory and voluntary approaches. Viewpoint. *Aquatic Conservation*, **3**, 269–280.

Gubbay, S. (1993b) A review of marine nature conservation initiatives in 1993 with special reference to the EC Habitats Directive (92/43/EEC), in *Marine Environmental Management. Review of Events in 1993 and Future Trends*, (ed. R. Earll), Vol. 1, Paper No. 5, 16–18.

Thompsett, P. and Turk, S. (1990) Helford River voluntary marine conservation area. *Marine Conservation*, **2**, 14, Marine Conservation Society, Ross-on-Wye.

Welton, S. (in press) *Caring for the Coast – A voluntary approach*, English Nature, Peterborough.

Wilson, N. (1991) *The effectiveness of Voluntary Marine Nature Reserves*, Freshwater and Marine Development Unit, Nature Conservancy Council, Peterborough (unpublished).

Index